Introduction to the theory of distributions

Introduction to the theory of distributions

F. G. FRIEDLANDER

Department of Pure Mathematics and Mathematical Statistics
University of Cambridge

with additional material by

M. Joshi

Department of Pure Mathematics and Mathematical Statistics
University of Cambridge

CAMBRIDGE
UNIVERSITY PRESS

PUBLISHED BY THE PRESS SYNDICATE OF THE UNIVERSITY OF CAMBRIDGE
The Pitt Building, Trumpington Street, Cambridge CB2 1RP, United Kingdom

CAMBRIDGE UNIVERSITY PRESS
The Edinburgh Building, Cambridge CB2 2RU, UK http://www.cup.cam.ac.uk
40 West 20th Street, New York, NY 10011-4211, USA http://www.cup.org
10 Stamford Road, Oakleigh, Melbourne 3166, Australia

First published 1982
Second edition 1998

Typeset in Times [VN]

A catalogue record for this book is available from the British Library

ISBN 0 521 64015 6 hardback
ISBN 0 521 64971 4 paperback

Transferred to digital printing 2003

CONTENTS

PREFACE

The aim of this book is to give a straightforward, and largely self-contained, introduction to the theory of distributions. It is based on lectures given in Cambridge over a number of years, to audiences who could not be expected to have much familiarity with such prerequisites as functional analysis, or Lebesgue integration. Distributions are therefore defined by means of semi-norm estimates; the relation of these to the underlying topological vector space theory is sketched in an appendix. Standard theorems on integrals which are needed are stated without proof. Such a presentation entails some loss, but this is offset, one hopes, by its wider accessibility. An interested reader can deepen his or her understanding by further study.

The first eight chapters of the book provide a basic course on the subject. The last two, which are independent of each other, carry the theory of Fourier transforms of distributions a little further. The theory is developed from the outset on open sets of $\mathbf{R}^n$, as the field in which distributions have been applied most conspicuously and fruitfully is the theory of partial differential equations. Some of these applications are included, often as exercises.

My debt to the literature, and in particular to L. Schwartz's treatise, and L. Hörmander's account of the subject in the first chapter of his Linear Partial Differential Operators, is apparent. In addition, I have been able to consult a set of unpublished lecture notes by Professor Hörmander. I also want to thank Richard Melrose, who has read the book in manuscript, and at the same time absolve him from any errors, for which I must take sole responsibility.

PREFACE TO THE SECOND EDITION

For this edition, a section on wave front sets has been added. This has been written by Dr Mark Joshi, and I want to thank him for his contribution. I have also used the opportunity to make some minor revisions and corrections. I am grateful to Ben Fairfax and Aurelian Bejancu, who have provided me with a comprehensive list of misprints, and raised some queries that have, I hope, led to improvements in the text.

Gerard Friedlander

INTRODUCTION

The theory of distributions is a generalization of classical analysis, which makes it possible to deal in a systematic manner with difficulties which previously had been overcome by *ad hoc* constructions, or just by pure hand waving. In fact, it does a good deal more: it provides a new and wider framework, and a more perspicuous language, in which one can reformulate and develop classical problems. Its influence has been particularly pervasive and fruitful in the theory of linear partial differential equations.

Let us consider some examples. If $(x, t) \in \mathbf{R}^2$, then

$$u = f(x+t) + g(x-t)$$

satisfies d'Alembert's equation

$$\frac{\partial^2 u}{\partial t^2} - \frac{\partial^2 u}{\partial x^2} = 0,$$

provided that the functions f and g are twice differentiable. This restriction is both irksome and unnatural in many instances. It can be overcome by introducing so-called weak solutions. By definition, these are functions u such that

$$\int u \left(\frac{\partial^2 \phi}{\partial t^2} - \frac{\partial^2 \phi}{\partial x^2} \right) \mathrm{d}x \, \mathrm{d}t = 0$$

for all sufficiently 'good' functions ϕ, for example for $\phi \in C_c^2(\mathbf{R}^2)$, the class of twice continuously differentiable functions that vanish on the exterior of a bounded set.

Again, if $x \in \mathbf{R}^3$, then the Newtonian potential

$$u(x) = \int \frac{f(y)}{|x-y|} \, \mathrm{d}y$$

satisfies Poisson's equation

$$\Delta u = -4\pi f$$

if the density function f is, for example, continuously differentiable. But it may not possess second order derivatives when f is merely continuous. Yet, it is then continuously differentiable and obeys Gauss's law that the flux of the field across a closed surface S is proportional to the matter enclosed by S. This difficulty can also be avoided by working with weak solutions of Poisson's equation. Furthermore, if one replaces f by the Dirac delta 'function', one obtains $u = 1/|x| = (x_1^2 + x_2^2 + x_3^2)^{-1/2}$ when $x \neq 0$, which is the potential due to a particle at the origin. It is suggestive to express this by writing

$$\Delta(1/|x|) = -4\pi\delta(x).$$

But this has to be interpreted; for example, one can take it to mean that

$$\int \frac{\Delta\phi(x)}{|x|} \,\mathrm{d}x = -4\pi\phi(0)$$

for all 'good' functions, say for all $\phi \in C_c^2(\mathbf{R}^3)$.

In all these cases, the difficulties and ambiguities disappear when the equations are read in terms of distributions. In the theory of distributions, functions are replaced by linear forms on an auxiliary vector space, whose members are called test functions. Recall that, if V is a vector space over the field $\mathbf{C}$ of complex numbers, then a linear form on V is a homomorphism $V \to \mathbf{C}$. The linear forms on V are made into a vector space $\mathrm{Hom}(V, \mathbf{C})$ in the obvious way: $\langle cu, \phi\rangle = c\langle u, \phi\rangle$ if $c \in \mathbf{C}$ and $u \in \mathrm{Hom}(V, \mathbf{C})$, and $\langle u + v, \phi\rangle = \langle u, \phi\rangle + \langle v, \phi\rangle$ if $u, v \in \mathrm{Hom}(V, \mathbf{C})$, where $\phi \in V$ in each case. In distribution theory, the basic space of test functions is $C_c^\infty(\mathbf{R}^n)$; its members are (complex valued) functions on $\mathbf{R}^n$ which possess continuous derivatives of all orders, and vanish outside some bounded set. The notations $C_0^\infty(\mathbf{R}^n)$, and L. Schwartz's original $\mathscr{D}(\mathbf{R}^n)$, are also used.

A continuous function f: $\mathbf{R}^n \to \mathbf{C}$ determines a linear form on $C_c^\infty(\mathbf{R}^n)$ by the rule

$$\langle f, \phi\rangle = \int_{\mathbf{R}^n} f\phi \,\mathrm{d}x, \quad \phi \in C_c^\infty(\mathbf{R}^n). \tag{1}$$

Conversely, it can (and will) be shown that this linear form determines f uniquely so that the space of continuous functions on $\mathbf{R}^n$ can be identified with a subspace of $\mathrm{Hom}(C_c^\infty(\mathbf{R}^n), \mathbf{C})$. If the function f is also continuously differentiable, then the linear forms on $C_c^\infty(\mathbf{R}^n)$ determined by its derivatives are, by (1) and an integration by parts,

$$\langle \partial f/\partial x_i, \phi\rangle = \int \phi(\partial f/\partial x_i) \,\mathrm{d}x = -\int f(\partial\phi/\partial x_i) \,\mathrm{d}x,$$

$$i = 1, \ldots, n, \ \phi \in C_c^\infty(\mathbf{R}^n).$$

Thus, for $i = 1, \ldots, n$,

$$\langle \partial f/\partial x_i, \phi\rangle = -\langle f, \partial\phi/\partial x_i\rangle, \quad \phi \in C_c^\infty(\mathbf{R}^n). \tag{2}$$

But this makes sense for any linear form on $C_c^\infty(\mathbf{R}^n)$, and so provides a definition of the derivatives of such a form. As one can iterate (2), one thus obtains (generalized) derivatives of all orders.

Multiplication by smooth (infinitely differentiable) functions can also be defined by analogy with the special case (1); one simply puts

$$\langle fu, \phi\rangle = \langle u, f\phi\rangle, \quad \phi \in C_c^\infty(\mathbf{R}^n). \tag{3}$$

By combining (2) and (3), one can thus account for the action of any linear differential operator with smooth coefficients on $\mathrm{Hom}(C_c^\infty(\mathbf{R}^n), \mathbf{C})$.

There is one other essential ingredient. The class of distributions is not the whole of $\mathrm{Hom}(C_c^\infty(\mathbf{R}^n), \mathbf{C})$: it is the subspace consisting of continuous linear forms. To say this, presupposes that $C_c^\infty(\mathbf{R}^n)$ has been equipped with an appropriate topology. The choice of this topology is, in fact, a cardinal feature of the theory of distributions. To define it, and to explore its implications, one must appeal to the theory of locally convex topological vector spaces. However, the course adopted in this book is to specify a certain set of inequalities which a linear form on $C_c^\infty(\mathbf{R}^n)$ must satisfy in order to qualify as a distribution. Once these are granted, the theory can be built up systematically and logically. But, so as to give some idea of what is involved, to readers who either do not have the time, or the inclination, to go into this more fully, a sketch of the functional-analytic background is given as an Appendix at the end of the book. This can be omitted, and the reader who does so should also ignore references to topological vector spaces in the text.

1 TEST FUNCTIONS AND DISTRIBUTIONS

1.1. Some notations and definitions

Throughout this book, the letter **R** will denote both the field of real numbers and the real line, and the letter **C** will stand for both the field of complex numbers and the complex plane. $\mathbf{R}^n$ will be understood to have its usual vector space structure and inner product, so that

$$cx = (cx_1, \ldots, cx_n) \quad \text{if} \quad c \in \mathbf{R} \quad \text{and} \quad x = (x_1, \ldots, x_n) \in \mathbf{R}^n,$$

and

$$x + y = (x_1 + y_1, \ldots, x_n + y_n), \quad x \cdot y = \sum_{j=1}^{n} x_j y_j$$

if $x \in \mathbf{R}^n$ and $y \in \mathbf{R}^n$. The Euclidean norm $(x \cdot x)^{1/2}$ will be written as $|x|$. We always write **R** instead of $\mathbf{R}^1$.

If A is a subset of $\mathbf{R}^n$, then $\bar{A}$ is its closure and ∂A is its boundary. It will be recalled that $A \subset \mathbf{R}^n$ is compact if and only if it is both closed and bounded; if $X \subset \mathbf{R}^n$ is an open set then A is a compact subset of X if it is compact and $A \subset X$. If A and B are subsets of $\mathbf{R}^n$, we shall write $A \setminus B$ for the (set-theoretic) difference $\{x : x \in A, x \notin B\}$.

If X and Y are sets, and f is a function on X with range in Y, one writes f: $X \to Y$, and $x \mapsto y$ or $x \mapsto f(x)$ indicates that f maps $x \in X$ to $y = f(x) \in Y$. A map is injective if $f(x') = f(x'')$ implies that $x' = x''$, surjective if every $y \in Y$ is the image of some $x \in X$, and bijective if it is both injective and surjective.

Let $X \subset \mathbf{R}^n$ be an open set, and let k be a nonnegative integer. The class $C^k(X)$ consists of the complex valued functions on X which have continuous derivatives of order less than, or equal to, k. (The function itself is included, conventionally, as the 'derivative of order zero'.) Likewise, $C^\infty(X)$ consists of the functions which have continuous derivatives of all orders.

The *support* of a function f: $X \to \mathbf{C}$ is the closure of the set $\{x \in X : f(x) \neq 0\}$; note that it is a closed subset of X. We shall write supp f for this. Functions with compact support play an important part in the theory. We write $C_c^k(X)$ for the subset of $C^k(X)$ consisting of functions with compact support, and

$C_c^\infty(X)$ for the subset of $C^\infty(X)$ consisting of functions with compact support. Note that $C^k(X)$, $C^\infty(X)$, $C_c^k(X)$ and $C_c^\infty(X)$ are all vector spaces over $\mathbf{C}$.

Integrals are Lebesgue integrals. When dealing with functions defined on a fixed open subset X of $\mathbf{R}^n$, we omit the domain of integration, so that

$$\int f(x)\, dx = \int_X f(x)\, dx$$

by definition; here, dx is Lebesgue measure.

Let $X \subset \mathbf{R}^n$ be an open set, and $f \in C^\infty(X)$. We shall usually write the derivatives as

$$\partial_j f = \partial f/\partial x_j, \quad j = 1, \ldots, n. \tag{1.1.1}$$

Derivatives of higher order can be written concisely by means of the multi-index notation. A *multi-index* (or, to be precise, an n-multi-index) is an n-tuplet $\alpha = (\alpha_1, \ldots, \alpha_n)$ of nonnegative integers; its length (or order) is $|\alpha| = \alpha_1 + \cdots + \alpha_n$. The sum of two multi-indices α and β is $\alpha + \beta = (\alpha_1 + \beta_1, \ldots, \alpha_n + \beta_n)$. One says that $\beta \leqslant \alpha$ if $\beta_j \leqslant \alpha_j$ for $j = 1, \ldots, n$; when $\beta \leqslant \alpha$ one can also define $\alpha - \beta = (\alpha_1 - \beta_1, \ldots, \alpha_n - \beta_n)$. One now sets

$$\partial^\alpha f = \partial^{|\alpha|} f/\partial x_1^{\alpha_1} \ldots \partial x_n^{\alpha_n}, \tag{1.1.2}$$

so that

$$\partial^\alpha \partial^\beta f = \partial^{\alpha+\beta} f. \tag{1.1.3}$$

This obviously also applies to $f \in C^k(X)$, provided that $|\alpha + \beta| \leqslant k$.

To complete the multi-index notation, we put

$$\alpha! = \alpha_1! \ldots \alpha_n! \tag{1.1.4}$$

for any multi-index α, and if also $x \in \mathbf{R}^n$ we set

$$x^\alpha = x_1^{\alpha_1} \ldots x_n^{\alpha_n}. \tag{1.1.5}$$

The formal statement of Taylor's theorem then becomes

$$f(x+h) = \sum_{\alpha \geqslant 0} \frac{x^\alpha}{\alpha!} \partial^\alpha f(h), \tag{1.1.6}$$

and the multinomial theorem assumes the concise form

$$(x_1 + \cdots + x_n)^m = \sum_{|\alpha| = m} \frac{m!}{\alpha!} x^\alpha. \tag{1.1.7}$$

1.2. Test functions

Let $X \subset \mathbf{R}^n$ be an open set. The members of the vector space $C_c^\infty(X)$ are called test functions. To give an example, we note first that, if $t \in \mathbf{R}$, the function

$$\psi_0(t) = e^{1/t} \quad \text{if} \quad t < 0, \quad \psi_0(t) = 0 \quad \text{if} \quad t \geqslant 0 \tag{1.2.1}$$

is a member of $C^\infty(\mathbf{R})$; the easy proof is left to the reader. Now put

$$\psi(x) = \psi_0(|x|^2 - 1). \tag{1.2.2}$$

Then $\psi \in C_c^\infty(\mathbf{R}^n)$, and the support of ψ is the closed unit ball $\{|x| \leqslant 1\}$. Other test functions can then be constructed by *regularization*; this is a method of 'smoothing' that will be extended to distributions in Chapter 4. For the present, we prove:

Theorem 1.2.1. Let $f \in C_c^k(\mathbf{R}^n)$, where $0 \leqslant k < \infty$. Let $\rho \in C_c^\infty(\mathbf{R}^n)$ be such that

$$\rho \geqslant 0, \quad \operatorname{supp} \rho \subset \{|x| \leqslant 1\}, \quad \int \rho \, dx = 1, \tag{1.2.3}$$

let ϵ be a positive real number, and put

$$f_\epsilon(x) = \epsilon^{-n} \int f(y) \rho\left(\frac{x-y}{\epsilon}\right) dy. \tag{1.2.4}$$

Then $f_\epsilon \in C_c^\infty(\mathbf{R}^n)$, and the support of f_ϵ is contained in the ϵ-neighbourhood of the support of f; moreover, if $|\alpha| \leqslant k$, then the $\partial^\alpha f_\epsilon$ converge uniformly to $\partial^\alpha f$ as $\epsilon \to 0$.

Note. One can, for example, take $\rho = \psi / \int \psi \, dx$ where ψ is given by (1.2.2).

Proof. The first statement is obvious, since repeated differentiation under the integral sign is permissible, and $f_\epsilon = 0$ when the distance of x from $\operatorname{supp} f$ exceeds ϵ. To prove the second assertion, write (1.2.4) as

$$f_\epsilon(x) = \int f(x - \epsilon z) \rho(z) \, dz. \tag{1.2.5}$$

By (1.2.3) one then has

$$|f_\epsilon(x) - f(x)| = \left| \int (f(x - \epsilon z) - f(x)) \rho(z) \, dz \right|$$

$$\leqslant \int |f(x - \epsilon z) - f(x)| \, \rho(z) \, dz$$

whence, again by (1.2.3),

$$|f_\epsilon(x) - f(x)| \leqslant \sup \{|f(x + y) - f(x)| : |y| \leqslant \epsilon\}.$$

This tends to zero uniformly as $\epsilon \to 0$, by the uniform continuity of f. If $|\alpha| \leqslant k$, one can differentiate under the integral sign in (1.2.5) to obtain

$$\partial^\alpha f_\epsilon(x) = \int \partial^\alpha f(x - \epsilon z) \rho(z) \, dz$$

and the same argument shows that $\partial^\alpha f_\epsilon \to \partial^\alpha f$ as $\epsilon \to 0$, with uniform convergence, and so we are done.

Remark. If $X \subset \mathbf{R}^n$ is an open set and $f \in C_c^k(X)$, then one can extend f to $\mathbf{R}^n$ by setting $f = 0$ on $\mathbf{R}^n \setminus \bar{X}$. Since $f_\epsilon \in C_c^\infty(X)$ when $\epsilon < \inf\{|x-y| : x \in \operatorname{supp} f\}$, $y \in \mathbf{R}^n \setminus X$ it is apparent that f can be approximated by test functions. Indeed, one can construct in this way a sequence $(f_j)_{1 \leqslant j < \infty} \in C_c^\infty(X)$ such that $\partial^\alpha f_j \to \partial^\alpha f$ as $j \to \infty$ if $|\alpha| \leqslant k$, the convergence being uniform.

The hypothesis that $\rho \geqslant 0$ will be useful in some applications but can obviously be dropped without invalidating Theorem 1.2.1.

1.3. Distributions

We are now ready to introduce distributions.

Definition 1.3.1. Let $X \subset \mathbf{R}^n$ be an open set. A linear form $u: C_c^\infty(X) \to \mathbf{C}$ is called a distribution if, for every compact set $K \subset X$, there is a real number $C \geqslant 0$ and a nonnegative integer N such that

$$|\langle u, \phi \rangle| \leqslant C \sum_{|\alpha| \leqslant N} \sup |\partial^\alpha \phi|, \tag{1.3.1}$$

for all $\phi \in C_c^\infty(X)$ with $\operatorname{supp} \phi \subset K$. The vector space of distributions on X is called $\mathscr{D}'(X)$.

Note. Inequalities such as (1.3.1) are called *semi-norm estimates*; the reason for this is explained in the Appendix.

The vector space structure of $\mathscr{D}'(X)$ is defined in the obvious way:

$$\langle u + v, \phi \rangle = \langle u, \phi \rangle + \langle v, \phi \rangle, \quad u, v \in \mathscr{D}'(X), \quad \phi \in C_c^\infty(X),$$

$$\langle cu, \phi \rangle = c\langle u, \phi \rangle, \quad c \in \mathbf{C}, \quad u \in \mathscr{D}'(X), \quad \phi \in C_c^\infty(X).$$

Our first example is simple, but important.

Theorem 1.3.1. Let $X \subset \mathbf{R}^n$ be an open set, and let $f \in C^0(X)$. Then

$$\langle f, \phi \rangle = \int f\phi \, dx, \quad \phi \in C_c^\infty(X) \tag{1.3.2}$$

is a distribution. Furthermore, if the second member of (1.3.2) vanishes for all $\phi \in C_c^\infty(X)$, then $f = 0$ on X.

Proof. That (1.3.2) is a distribution is obvious, since

$$\left| \int f\phi \, dx \right| \leqslant \sup |\phi| \int_K |f| \, dx, \quad \phi \in C_c^\infty(K), \tag{1.3.3}$$

where $K \subset X$ is a compact set and

$$C_c^\infty(K) = \{\phi : \phi \in C_c^\infty(\mathbf{R}^n), \quad \operatorname{supp} \phi \subset K\}. \tag{1.3.4}$$

Suppose now that the second member of (1.3.2) vanishes for all $\phi \in C_c^\infty(X)$, but that $f(y) \neq 0$ at some point $y \in X$. By the continuity of f, there is then a $\delta > 0$ such that $\operatorname{Re}(f(x)/f(y)) \geqslant \frac{1}{2}$ if $|x-y| \leqslant \delta$, and one can clearly assume that $\{x: |x-y| \leqslant \delta\} \subset X$. Hence, with a real valued $\rho \in C_c^\infty(\mathbf{R}^n)$ satisfying (1.2.3), one has

$$0 = \operatorname{Re} \int \frac{f(x)}{f(y)}\, \rho\left(\frac{x-y}{\delta}\right) \mathrm{d}x \geqslant \tfrac{1}{2} \int \rho\left(\frac{x-y}{\delta}\right) \mathrm{d}x = \tfrac{1}{2}\delta^n,$$

which is absurd. So the theorem is proved.

Thus, (1.3.2) gives an injective map $C^0(X) \to \mathscr{D}'(X)$, and one can identify a continuous function f with the distribution defined in this way. We shall usually speak of the distribution determined by f, or of the distribution equal to f.

Remark. It is obvious from (1.3.3) that (1.3.2) also yields a distribution on X when f is *locally integrable*, that is to say when (it is measurable and) $\int_K |f|\, \mathrm{d}x < \infty$ for all compact sets $K \subset X$. This distribution cannot determine f uniquely, as the second member of (1.3.2) is unchanged when f is replaced by a function g that is equal to f almost everywhere. (This means that $\{x \in X: f(x) - g(x) \neq 0\}$ is a set of measure zero. A set of measure zero is one that can, given any $\epsilon > 0$, be covered by a countable family of rectangles whose total volume is less than ϵ.) However, it is not difficult to show that, if f is locally integrable, and the second member of (1.3.2) vanishes for all $\phi \in C_c^\infty(X)$, then $f = 0$ almost everywhere. (The reader familiar with integration will be able to deduce this from Theorem 1.2.1 by considering the measure $f\, \mathrm{d}x$.) By definition, the vector space $L_1^{\text{loc}}(X)$ consists of all equivalence classes of locally integrable functions on X. So (1.3.2) gives an injective map $L_1^{\text{loc}}(X) \to \mathscr{D}'(X)$; we shall simply refer to the distribution determined by (or 'equal to') f.

Another basic example is the *Dirac distribution* (delta function) which is the member of $\mathscr{D}'(\mathbf{R}^n)$ given by

$$\langle \delta, \phi \rangle = \phi(0), \quad \phi \in C_c^\infty(\mathbf{R}^n). \tag{1.3.5}$$

More generally, if $X \subset \mathbf{R}^n$ is an open set and y is a point of X then $\delta_y \in \mathscr{D}'(X)$ is defined by

$$\langle \delta_y, \phi \rangle = \phi(y), \quad \phi \in C_c^\infty(X). \tag{1.3.6}$$

It is easy to show that neither (1.3.5) nor (1.3.6) can be put into the form (1.3.2); the proof is left to the reader.

There is another way of characterizing distributions. For this, we need a definition.

Definition 1.3.2. Let $X \subset \mathbf{R}^n$ be an open set. A sequence $(\phi_j)_{1 \leqslant j < \infty} \in C_c^\infty(X)$ is said to converge (or tend) to zero in $C_c^\infty(X)$ if (i) the supports of the ϕ_j are contained in a fixed compact subset of X, and (ii) for each multi-index α, the $\partial^\alpha \phi_j$ converge to zero uniformly as $j \to \infty$.

Theorem 1.3.2. A linear form u on $C_c^\infty(X)$ is a distribution if and only if $\lim_{j\to\infty}\langle u, \phi_j\rangle = 0$ for every sequence ϕ_j which converges to zero in $C_c^\infty(X)$ as $j \to \infty$.

Note. This property is called *sequential continuity.*

Proof. It is obvious from Definitions 1.3.1 and 1.3.2 that the condition is necessary. To prove sufficiency, one argues by contradiction. Suppose, then, that u is sequentially continuous, but is not a distribution. Then there is a compact set $K \subset X$ such that, for each nonnegative integer N, the numbers

$$|\langle u, \phi\rangle| / \sum_{|\alpha| \leqslant N} \sup |\partial^\alpha \phi|, \quad \phi \in C_c^\infty(K)$$

are an unbounded subset of $[0, \infty)$. (For the definition of $C_c^\infty(K)$, see (1.3.4).) Hence for each $N = 0, 1, \ldots$, there is a $\phi_N \in C_c^\infty(K)$ such that

$$|\langle u, \phi_N\rangle| \geqslant N \sum_{|\alpha| \leqslant N} \sup |\partial^\alpha \phi_N|. \tag{1.3.7}$$

Put

$$\psi_N(x) = \phi_N(x) / \sum_{|\alpha| \leqslant N} \sup |\partial^\alpha \phi_N|.$$

Then $\operatorname{supp} \psi_N \subset K$, and $|\partial^\beta \psi_N| \leqslant 1/N$ for $|\beta| \leqslant N$. Hence $\psi_N \to 0$ in $C_c^\infty(X)$ as $N \to \infty$. But it follows from (1.3.7) that $|\langle u, \psi_N\rangle| \geqslant 1$ for all N. So we have arrived at a contradiction, and the proof is complete.

At this point, one can introduce a useful subspace of $\mathscr{D}'(X)$, the *distributions of finite order*.

Definition 1.3.3. A distribution $u \in \mathscr{D}'(X)$ is said to be of finite order if one can take the same N in (1.3.1) for all compact sets $K \subset X$; its order is then the least such N. The vector space of distributions of order $\leqslant m$ is called $\mathscr{D}'^m(X)$.

One can also modify Definition 1.3.2 as follows:

Definition 1.3.2′. A sequence $(\phi_j)_{1 \leqslant j < \infty} \in C_c^m(X)$ is said to converge to zero in $C_c^m(X)$ if the supports of all the ϕ_j are contained in a fixed compact subset of X, and the $\partial^\alpha \phi_j$, where $|\alpha| \leqslant m$, converge uniformly to zero as $j \to \infty$.

One then has

Theorem 1.3.3. A distribution $u \in \mathscr{D}'^m(X)$, where $0 \leqslant m < \infty$, has a unique extension to a linear form on $C_c^m(X)$ which is sequentially continuous. Conversely, the restriction of a sequentially continuous linear form on $C_c^m(X)$ to $C_c^\infty(X)$ is a member of $\mathscr{D}'^m(X)$.

Proof. If $u \in \mathscr{D}'^m(X)$, then one has semi-norm estimates

$$|\langle u, \phi\rangle| \leqslant C(K) \sum_{|\alpha| \leqslant m} \sup |\partial^\alpha \phi|, \quad \phi \in C_c^\infty(K). \tag{1.3.8}$$

It follows from Theorem 1.2.1 that if $\phi \in C_c^m(X)$ is given, then one can construct a sequence $(\phi_j)_{1 \leqslant j < \infty} \in C_c^\infty(X)$ which converges to ϕ in $C_c^m(X)$ as $j \to \infty$. It is clear from (1.3.8) that the numerical sequence $\langle u, \phi_j\rangle$ converges as $j \to \infty$, and that its limit provides a linear form on $C_c^m(X)$ that satisfies (1.3.8) for all $\phi \in C_c^m(X)$; it is also evident that this extension is unique. The second assertion follows from Theorem 1.3.2; the details are left to the reader.

Both our examples, (1.3.2) and (1.3.5), are distributions of order zero. An example of a distribution of order $m > 0$ is

$$\phi \mapsto \partial^\alpha \phi(0), \quad \phi \in C_c^\infty(\mathbf{R}^n), \quad |\alpha| = m.$$

Remark. A complex, locally finite Borel measure μ defined on an open set $X \subset \mathbf{R}^n$ determines a distribution by

$$\langle \mu, \phi\rangle = \int \phi \, \mathrm{d}\mu, \quad \phi \in C_c^\infty(X). \tag{1.3.9}$$

Conversely, it follows from Theorem 1.2.1 and the Riesz representation theorem (see [5, p. 39] or [8, p. 119]) that every distribution of order 0 is of the form (1.3.9). The Dirac distribution (1.3.5) is of course a particular instance, and is therefore often called *Dirac measure*.

1.4. Localization

If $X \subset \mathbf{R}^n$ is an open set and X' is an open subset of X, then one has an inclusion $C_c^\infty(X') \to C_c^\infty(X)$, since one can extend any $\phi \in C_c^\infty(X')$ to X by setting $\phi = 0$ on $X \setminus \bar{X}'$. So, if $u \in \mathscr{D}'(X)$, then

$$\phi \mapsto \langle u, \phi\rangle, \quad \phi \in C_c^\infty(X')$$

is a distribution on X'; the semi-norm estimates (1.3.1) yield the requisite semi-norm estimates on X'. This is the *restriction* of u to X'. Two distributions u and v defined on X are then said to be *equal on* X' if their restrictions are equal, that is to say if $\langle u, \phi\rangle = \langle v, \phi\rangle$ for all $\phi \in C_c^\infty(X')$.

The restriction of $u \in \mathscr{D}'(X)$ to an open subset of X is also called a *localization*. A distribution can be recovered from its localizations. To prove this, one needs an important technical device, called a partition of unity. The next theorem is a simple version of this, which will be sufficient for our purpose.

Theorem 1.4.1. Let $X \subset \mathbf{R}^n$ be an open set, and let K be a compact subset of X. Let $X_i, i=1,\ldots,m$, be open subsets of X whose union contains K. Then one can find functions $\psi_i, \ldots, \psi_m$ such that

$$\psi_i \in C_c^\infty(X_i), \quad 0 \leqslant \psi_i \leqslant 1, \quad i=1,\ldots,m,$$

$$\sum_{i=1}^m \psi_i \leqslant 1 \quad \text{on } X, \quad \sum_{i=1}^m \psi_i = 1 \quad \text{on a neighbourhood of } K. \quad (1.4.1)$$

Proof. Suppose first that $m=1$. The distance $d(x,A)$ of a point $x \in \mathbf{R}^n$ from a set $A \subset \mathbf{R}^n$ is

$$d(x,A) = \inf\{|x-y| : y \in A\}. \quad (1.4.2)$$

It follows from the triangle inequality that $x \mapsto d(x,A)$ is continuous. For any $\delta > 0$, set $K_\delta = \{x : d(x,K) < \delta\}$; this is the δ-neighbourhood of K. Let ϵ be a positive real number, chosen so that

$$4\epsilon < \inf\{|x-y| : x \in K, \quad y \in \mathbf{R}^n \setminus X_1\}$$

and put

$$f(x) = 1 - \epsilon^{-1} d(x, K_{2\epsilon}), \quad x \in K_{3\epsilon},$$
$$= 0, \quad x \notin K_{3\epsilon}.$$

With this f, the second member of (1.2.4) yields a function $\psi \in C_c^\infty(X_1)$ such that $\psi = 1$ on K_ϵ, as required.

Suppose now that $m>1$. For each $x \in K$, one can choose a positive real number $r=r(x)$ such that the closure of the ball $B(x;r) = \{x' : |x'-x| < r\}$ is contained in an X_i. Thus one gets an open cover $\{B(x;r) : x \in K\}$ of K, and as K is compact this contains a finite subcover $(B_1, \ldots, B_N)$. For each $i, i=1,\ldots,m$ let K_i be the union of closed balls $\bar{B}_j$ which are contained in X_i. Then each K_i is a compact subset of the corresponding X_i, and K is contained in the union of the K_i. By the result for $m=1$ which has just been established, one can find functions $\phi_i \in C_c^\infty(X_i)$ such that $0 \leqslant \phi_i \leqslant 1$ and $\phi_i = 1$ on a neighbourhood of K_i, for each $i=1,\ldots,m$; extend these trivially to elements of $C_c^\infty(X)$. Finally, put

$$\psi_1 = \phi_1, \quad \psi_2 = \phi_2(1-\phi_1), \ldots, \quad \psi_m = \phi_m(1-\phi_1)\ldots(1-\phi_{m-1}).$$

Then supp $\psi_i \subset X_i$, and $0 \leqslant \psi_i \leqslant 1$, for $i=1,\ldots,m$ and

$$\sum_{i=1}^m \psi_i = 1 - (1-\phi_1)\ldots(1-\phi_m)$$

So it is clear that (1.4.1) holds, and we are done.

Remark. If $K \subset \mathbf{R}^n$ is a compact set, and X is an open neighbourhood of K, then Theorem 1.4.1 (with $m=1$) shows that one can find a $\psi \in C_c^\infty(\mathbf{R}^n)$ such that supp $\psi \subset X$, $0 \leqslant \psi \leqslant 1$, and $\psi = 1$ on a neighbourhood of K. Such a function will be called a *cut-off function*. (It is also known as a 'bump function'.)

As a first application of this theorem, we discuss the support of a distribution:

Definition 1.4.1. Let $X \subset \mathbf{R}^n$ be an open set, and let $u \in \mathscr{D}'(X)$. The support of u, written as supp u, is the complement of the set

$$\{x: u = 0 \text{ on a neighbourhood of } x\}$$

Note that this is a closed subset of X. For example, the support of the Dirac distribution is the origin, and one has $\delta = 0$ on $\mathbf{R}^n \setminus \{0\}$. This is a particular case of the next ('obvious') theorem.

Theorem 1.4.2. Let $u \in \mathscr{D}'(X)$ and $\phi \in C_c^\infty(X)$. If the supports of u and of ϕ are disjoint, then $\langle u, \phi \rangle = 0$.

Proof. Put $K = \operatorname{supp} \phi$. By hypothesis, every point of the compact set K has an open neighbourhood on which $u = 0$. This open cover of K contains a finite subcover $X_1, \ldots, X_m$. With $\psi_1, \ldots, \psi_m$ as in (1.4.1) one then has

$$\langle u, \phi \rangle = \left\langle u, \sum_{i=1}^{m} \phi\psi_i \right\rangle = \sum_{i=1}^{m} \langle u, \phi\psi_i \rangle = 0,$$

since $\phi\psi_i \in C_c^\infty(X_i)$ and $u = 0$ on X_i for $i = 1, \ldots, m$ by hypothesis; this proves the theorem.

Corollary. If $u \in \mathscr{D}'(X)$ and every point of X has a neighbourhood on which $u = 0$, then $u = 0$.

More generally, a distribution can be recovered from its localizations.

Theorem 1.4.3. Let $X \subset \mathbf{R}^n$ be an open set, and let $(X_\lambda)_{\lambda \in \Lambda}$, where Λ is an index set, be an open cover of X. Suppose that, for each $\lambda \in \Lambda$, there is given a distribution $u_\lambda \in \mathscr{D}'(X_\lambda)$, and that

$$u_\lambda = u_\mu \quad \text{on} \quad X_\lambda \cap X_\mu \quad \text{if} \quad X_\lambda \cap X_\mu \neq \emptyset. \tag{1.4.3}$$

Then there is a unique $u \in \mathscr{D}'(X)$ such that $u = u_\lambda$ in X_λ, for each $\lambda \in \Lambda$.

Proof. We only have to prove existence, as uniqueness follows from the Corollary to Theorem 1.4.2. So, let $\phi \in C_c^\infty(X)$. Since the X_λ are an open cover of supp ϕ, there is a finite set $X_{\lambda(1)}, \ldots, X_{\lambda(m)}$ which is an open cover of supp ϕ. By Theorem 1.4.1, one can find functions $\psi_1, \ldots, \psi_m$ which satisfy (1.4.1) with $X_i = X_{\lambda(i)}$, $i = 1, \ldots, m$, and so one can set

$$\langle u, \phi \rangle = \sum_{i=1}^{m} \langle u_{\lambda(i)}, \phi\psi_i \rangle \tag{1.4.4}$$

Consider another such open cover $X'_{\mu(1)}, \ldots, X'_{\mu(l)}$ of supp ϕ, and a corresponding set of functions $\psi'_n, \ldots, \psi'_l$. Then

$$\sum_{i=1}^{m} \langle u_{\lambda(i)}, \phi\psi_i\rangle = \sum_{i=1}^{m}\sum_{j=1}^{l} \langle u_{\lambda(i)}, \phi\psi_i\psi_j'\rangle$$

$$= \sum_{i=1}^{m}\sum_{j=1}^{l} \langle u_{\mu(j)}, \phi\psi_i\psi_j'\rangle,$$

by (1.4.3). Summing over i gives

$$\sum_{i=1}^{m} \langle u_{\lambda(i)}, \phi\psi_i\rangle = \sum_{j=1}^{l} \langle u_{\mu(j)}, \phi\psi_j'\rangle.$$

Hence (1.4.4) defines $\langle u, \phi\rangle$ unambiguously. Moreover, if supp $\phi \subset X_\lambda$ for some $\lambda \in \Lambda$ then one can just take $l = 1$ and conclude that $u = u_\lambda$ in X_λ.

Finally, it is easy to see that (1.4.4) is a distribution. For if K is a compact subset of X, one can use the same set of ψ_i for all $\phi \in C_c^\infty(K)$, and Leibniz's theorem then shows that the linear form defined by (1.4.4) satisfies a semi-norm estimate. The simple details are left to the reader. So the theorem is proved.

Note on partitions of unity. Let $(X_\lambda)_{\lambda\in\Lambda}$ be an open cover of an open subset X of $\mathbf{R}^n$. By an argument similar to the proof of Theorem 1.4.4 one can show that there exists a sequence of functions $(\psi_i)_{1\leqslant j<\infty} \in C_c^\infty(X)$ such that: (i) each ψ_i has its support in some X_λ; (ii) one has

$$\psi_i(x) \geqslant 0, \quad \sum_{i=1}^{\infty} \psi_i(x) = 1, \quad x \in X; \tag{1.4.5}$$

(iii) any compact set intersects only a finite number of the supports of the ψ_i. Such a collection of functions is called a *locally finite partition of unity, subordinated to the cover* (X_λ). See [4, p. 147], [6, p. 22] or [8, p. 60].

1.5. Convergence of distributions

We conclude this Chapter by introducing the notion of a convergent sequence of distributions.

Definition 1.5.1. Let $X \subset \mathbf{R}^n$ be an open set, and let $(u_j)_{1\leqslant j<\infty}$ be a sequence of distributions on X. The sequence is said to converge in $\mathscr{D}'(X)$ to $u \in \mathscr{D}'(X)$ if

$$\lim \langle u_j, \phi\rangle = \langle u, \phi\rangle \quad \text{for all } \phi \in C_c^\infty(X). \tag{1.5.1}$$

Note. This definition of convergence extends in an obvious way to certain families of distributions depending on a continuous parameter. For instance, Theorem 1.2.1 shows that if $\rho \in C_c^\infty(\mathbf{R}^n)$ satisfies the hypotheses (1.2.3), then the functions $x \to \epsilon^{-n}\rho(x/\epsilon)$, where $\epsilon > 0$, converge as distributions to the Dirac distribution in $\mathscr{D}'(X)$ as $\epsilon \to 0+$.

When the u_j are distributions determined by functions, one must in general distinguish between convergence in the usual sense ('pointwise convergence'), and convergence in $\mathscr{D}'$. For example, if $X=\mathbf{R}$ and $u_m=\exp(imx), m=1,2,\ldots,$ then the u_m converge pointwise only when $x=2k\pi$, where k is a rational integer. But the identity

$$\langle u_m, \phi\rangle = \int \phi(x)\,\mathrm{e}^{imx}\,\mathrm{d}x = -\frac{1}{im}\int \phi'(x)\,\mathrm{e}^{imx}\,\mathrm{d}x, \quad \phi\in C_c^\infty(\mathbf{R}),$$

which is obtained by integration by parts, shows that $u_m\to 0$ in $\mathscr{D}'(\mathbf{R})$ as $m\to\infty$.

Again, suppose that $f\in C^0(\mathbf{R})$ has its support contained in $[0,1]$ and that

$$\int f(x)\,\mathrm{d}x = 1. \qquad (1.5.2)$$

Put

$$f_k(x) = kf(kx), \quad k=1,2,\ldots.$$

Then $f_k(x)\to 0$ for any fixed $x\in\mathbf{R}$ as $k\to\infty$. On the other hand, it follows from (1.5.2) that, if $\phi\in C_c^\infty(\mathbf{R})$, then

$$|\langle f_k,\phi\rangle - \phi(0)| \leqslant \left(\int |f(x)|\,\mathrm{d}x\right)\sup\{|\phi(x)-\phi(0)|: 0\leqslant x\leqslant 1/k\}.$$

Hence $f_k\to\delta$ in $\mathscr{D}'(\mathbf{R})$ as $k\to\infty$.

A useful criterion characterizing sequences of functions whose pointwise limit coincides with the limit in $\mathscr{D}'$ can be derived from the *dominated convergence theorem*. This is one of the basic theorems of (Lebesgue) integrals. It says that, if $(f_j)_{1\leqslant j<\infty}$ is a sequence of integrable functions which converges almost everywhere on an open set $X\subset\mathbf{R}^n$, and if there is an integrable nonnegative function g such that $|f_j|\leqslant g$ for all j, then $\lim_{j\to\infty}f_j$ is also integrable on X, and

$$\lim_{j\to\infty}\int_X f_j\,\mathrm{d}x = \int_X \lim_{j\to\infty} f_j\,\mathrm{d}x.$$

Theorem 1.5.1. If $(f_j)_{1\leqslant j<\infty}\in L_1^{\mathrm{loc}}(X)$ is a sequence which converges almost everywhere to a function f, and there is a function $g\in L_1^{\mathrm{loc}}(X)$ such that $|f_j|\leqslant g$ for all j, then $f\in L_1^{\mathrm{loc}}(X)$ and $f_j\to f$ in $\mathscr{D}'(X)$ as $j\to\infty$.

Proof. Apply the dominated convergence theorem to conclude that

$$\langle f_j,\phi\rangle = \int f_j\phi\,\mathrm{d}x \to \int f\phi\,\mathrm{d}x = \langle f,\phi\rangle \text{ as } j\to\infty,$$

for all $\phi\in C_c^\infty(X)$.

Note that this includes the elementary case of a sequence of continuous functions which converges uniformly on compact sets.

In Definition 1.5.1, the existence of the limit $u \in \mathscr{D}'(X)$ is postulated. In fact, one can say more:

Theorem 1.5.2. Let $X \subset \mathbf{R}^n$ be an open set, and let $(u_j)_{1 \leq j < \infty}$ be a sequence of distributions on X which has the property that, for each $\phi \in C_c^\infty(X)$, the sequence $\langle u_j, \phi \rangle$ converges as $j \to \infty$. Then

$$\phi \to \langle u, \phi \rangle = \lim_{j \to \infty} \langle u_j, \phi \rangle, \quad \phi \in C_c^\infty(X),$$

is a member of $\mathscr{D}'(X)$.

The proof of this will be omitted. It is a consequence of an important result in functional analysis, the Banach-Steinhaus theorem. See, for example, [4, pp. 43-5 and p. 146]. A direct proof can be found in [2] or in [9]. This property of $\mathscr{D}'$ ('sequential completeness') is important when one is dealing with distributions which are functions of a parameter. For example, suppose that $u_t \in \mathscr{D}'(X)$ is a family of distributions depending on a real variable t, and that $t \mapsto \langle u_t, \phi \rangle$ is differentiable for each $\phi \in C_c^\infty(X)$; then $\phi \mapsto (\partial/\partial t)\langle u_t, \phi \rangle, \phi \in C_c^\infty(X)$ is a distribution, which is denoted by $(\partial/\partial t)u_t$.

Exercises

1.1. Let $x \in \mathbf{R}^n$ and put $g_\alpha(x) = x^\alpha$, where α is a multi-index; compute $\partial^\beta g_\alpha(x)$ for all multi-indices $\beta > 0$.

Hence prove Taylor's theorem (1.1.6) when f is a polynomial, and deduce the multinomial theorem (1.1.7) from this.

1.2. Show that if $f \in C^0(\mathbf{R}^n)$, then its support is identical with the support of the distribution

$$\langle f, \phi \rangle = \int f\phi \, dx, \quad \phi \in C_c^\infty(\mathbf{R}^n).$$

Is this true when $f \in L_1^{\text{loc}}(\mathbf{R}^n)$?

1.3. Show that the principal value integral

$$\text{p.v.} \int \frac{\phi(x)}{x} \, dx = \lim_{\epsilon \to 0+} \left(\int_{-\infty}^{-\epsilon} \frac{\phi(x)}{x} \, dx + \int_{\epsilon}^{\infty} \frac{\phi(x)}{x} \, dx \right)$$

exists for all $\phi \in C_c^\infty(\mathbf{R})$, and is a distribution. What is its order?

1.4. Find a distribution $u \in \mathscr{D}'(\mathbf{R})$ such that $u = 1/x$ on $(0, \infty)$ and $u = 0$ on $(-\infty, 0)$.

1.5. Show that

$$\langle u, \phi \rangle = \sum_{k=1}^{\infty} \partial^k \phi(1/k)$$

is a distribution on $(0, \infty)$, but that there is no $v \in \mathscr{D}'(\mathbf{R})$ whose restriction to $(0, \infty)$ is equal to u.

1.6. Let $u \in \mathscr{D}'(\mathbf{R}^n)$ have the property that $\langle u, \phi \rangle \geqslant 0$ for all real valued nonnegative $\phi \in C_c^\infty(\mathbf{R}^n)$. Show that u is of order 0.

1.7. Let $f_\epsilon \in L_1^{\text{loc}}(\mathbf{R}^n)$ be a function which depends on a parameter $\epsilon \in (0, 1)$, and is such that

(*a*) $\operatorname{supp} f_\epsilon \subset \{|x| \leqslant \epsilon\}$,

$$(b) \int f_\epsilon(x)\, dx = 1,$$

$$(c) \int |f_\epsilon(x)|\, dx \leqslant \mu < \infty, \qquad 0 < \epsilon < 1.$$

Show that $f_\epsilon \to \delta$ in $\mathscr{D}'(\mathbf{R}^n)$ as $\epsilon \to 0$.

Show also that, if f_ϵ satisfies (*a*) and $f_\epsilon \to \delta$ in $\mathscr{D}'(\mathbf{R}^n)$ as $\epsilon \to 0$, then (*b*) holds.

1.8. Let $(f_k)_{1 \leqslant k < \infty} \in L_1^{\text{loc}}(\mathbf{R}^n)$ be a sequence of real valued functions such that

$$\operatorname{supp} f_k \subset \{|x| \leqslant k^{-1}\}, \qquad \int f_k(x)\, dx = 1, \qquad k = 1, 2, \ldots$$

Show that the sequence $(f_k^2)_{1 \leqslant k < \infty}$ does not converge in $\mathscr{D}'(\mathbf{R}^n)$ as $k \to \infty$.

1.9. Let $(c_k)_{k \in \mathbf{Z}}$ be complex numbers which satisfy

$$|c_k| \leqslant C(1 + |k|)^m, \qquad k \in \mathbf{Z},$$

for some constants $C \geqslant 0$ and m. Show that

$$u = \sum_{k \in \mathbf{Z}} C_k\, e^{ikx}$$

converges in $\mathscr{D}'(\mathbf{R})$. (**Z** stands for the rational integers.)

1.10. Construct a function $\phi \in C_c^\infty(\mathbf{R})$ such that $\phi \geqslant 0$, $\operatorname{supp} \phi \subset (-1, 1)$, and the functions

$$\psi_n(x) = \phi(x - n) / \sum_{m \in Z} \phi(x - m), \qquad n \in \mathbf{Z}$$

are a partition of unity subordinated to the cover $\cup_{n \in \mathbf{Z}} (n - 1, n + 1)$ of R.

2 DIFFERENTIATION, AND MULTIPLICATION BY SMOOTH FUNCTIONS

In this section, two basic operations on distributions are introduced; differentiation, and multiplication by C^∞ functions. Each is a continuous linear map $\mathscr{D}' \to \mathscr{D}'$; by combining them, one obtains linear differential operators.

Throughout the section, X is an open set in $\mathbf{R}^n$.

2.1. The derivatives of a distribution

If $u \in C^1(X)$, then the distribution which is equal to the derivative $\partial_i u$, where $i = 1, \ldots, n$, is

$$\langle \partial_i u, \phi \rangle = \int \phi \partial_i u \, \mathrm{d}x = -\int u \partial_i \phi \, \mathrm{d}x, \quad \phi \in C_c^\infty(X), \tag{2.1.1}$$

by a partial integration. One can write this as

$$\langle \partial_i u, \phi \rangle = -\langle u, \partial_i \phi \rangle, \; i = 1, \ldots, n, \quad \phi \in C_c^\infty(X). \tag{2.1.2}$$

This also makes sense for any $u \in \mathscr{D}'(X)$. Linearity is obvious. Now $C_c^\infty(X)$ is stable under differentiation, since the derivatives of test functions are themselves test functions. Moreover, one has supp $\partial^\alpha \phi \subset$ supp ϕ for all α. So, if $K \subset X$ is a compact set, then the semi-norm estimate (1.3.1) implies that

$$|\langle u, \partial_i \phi \rangle| \leqslant C \sum_{|\alpha| \leqslant N+1} \sup |\partial^\alpha \phi|, \quad \phi \in C_c^\infty(K).$$

Thus the second members of (2.1.2) are distributions, and we can introduce them formally:

Definition 2.1.1. Let $u \in \mathscr{D}'(X)$. The distributions $\partial_i u, i = 1, \ldots, n$, given by (2.1.2), are called the first order derivatives of u.

A distribution has derivatives of all orders. For one can iterate (2.1.2) to obtain

$$\langle \partial^\alpha u, \phi \rangle = (-1)^{|\alpha|} \langle u, \partial^\alpha \phi \rangle, \quad \phi \in C_c^\infty(X). \tag{2.1.3}$$

Evidently, the map $\partial^\alpha : \mathscr{D}'(X) \to \mathscr{D}'(X)$ defined by $u \mapsto \partial^\alpha u$ is a vector space homomorphism. An immediate consequence of the definition is that this map is also sequentially continuous.

Theorem 2.1.1. If $(u_k)_{1\leq k<\infty} \in \mathscr{D}'(X)$ is a sequence which converges to u in $\mathscr{D}'(X)$ as $k\to\infty$, and α is a multi-index, then one also has $\partial^\alpha u_k \to \partial^\alpha u$ in $\mathscr{D}'(X)$ as $k\to\infty$.

Proof. It follows from (2.1.3) that as $k\to\infty$,

$$\langle \partial^\alpha u_k, \phi\rangle = (-1)^{|\alpha|}\langle u_k, \partial^\alpha \phi\rangle \to (-1)^{|\alpha|}\langle u, \partial^\alpha\phi\rangle = \langle \partial^\alpha u, \phi\rangle$$

for any $\phi\in C_c^\infty(X)$; so we are done.

Definition 2.1.1, the identity (2.1.1), and Theorem 1.3.1 show that the distributions $\partial_i u$ are equal to the usual first order derivatives of u when $u\in C^1(X)$. This hypothesis can be weakened.

Theorem 2.1.2. Let f and g be continuous functions on X, and suppose that they satisfy $\partial_i f = g$ as distributions. Then the usual partial derivative $\partial f/\partial x_i$ exists, and is equal to g.

Proof. Let $B(y;r)$ denote the ball $\{x: |x-y|<r\}$ for any point $y\in X$ and positive number r. Chose $\delta>0$ such that $B(y;2\delta)\subset X$. If $0<\epsilon<\delta$, then one can construct functions f_ϵ and g_ϵ by regularization as in the proof of Theorem 1.2.1, and both are in $C^\infty(B(y;\delta))$. The hypothesis gives $\partial f_\epsilon/\partial x_i = g_\epsilon$. The theorem now follows by uniform convergence if one makes $\epsilon\to 0$, as y can be any point of X. The details are left as an exercise.

2.2. Some examples

Let us first take $X=\mathbf{R}$. The *Heaviside function* is, by definition,

$$H(x)=1, \quad x>0; \quad H(x)=0, \quad x\leq 0. \tag{2.2.1}$$

To compute its derivative, one has

$$\langle \partial H, \phi\rangle = -\langle H, \partial\phi\rangle = -\int_0^\infty \partial\phi(x)\,dx = \phi(0), \quad \phi\in C_c^\infty(\mathbf{R}).$$

Hence

$$\partial H = \delta. \tag{2.2.2}$$

For a fixed $a\in R$ a similar manipulation gives

$$\partial_x H(x-a) = \delta_a(x) \tag{2.2.3}$$

By (2.1.3) the higher order derivatives of the Dirac distribution are

$$\langle \partial^k\delta_a, \phi\rangle = (-1)^k \partial^k\phi(a), \quad k=1,2,\ldots, \phi\in C_c^\infty(\mathbf{R}). \tag{2.2.4}$$

In the same manner, one can derive a rule for ‘differentiating a discontinuity’. If $f\in C^\infty(\mathbf{R})$ then $x\mapsto H(x)f(x)$ is locally integrable and so determines a distribution. Then, with $\phi\in C_c^\infty(\mathbf{R})$,

$$\langle \partial(fH), \phi \rangle = -\langle fH, \partial\phi \rangle = \int_0^\infty f(x)\partial\phi(x)\,\mathrm{d}x$$

$$= f(0)\phi(0) + \int_0^\infty \phi(x)\partial f(x)\,\mathrm{d}x,$$

whence

$$\partial(fH) = f(0)\delta + H\partial f. \tag{2.2.5}$$

It follows by induction that

$$\partial^m(fH) = \sum_{k=0}^{m-1} \partial^k f(0)\partial^{m-k-1}\delta + H\partial^m f, \tag{2.2.6}$$

for all positive integers m.

Next, we consider the *principal value* distribution x^{-1}. This is

$$\text{p.v.} \int \frac{\phi(x)}{x}\,\mathrm{d}x = \lim_{\epsilon\to 0+} \left(\int_{-\infty}^{-\epsilon} \frac{\phi(x)}{x}\,\mathrm{d}x + \int_{\epsilon}^{\infty} \frac{\phi(x)}{x}\,\mathrm{d}x \right), \tag{2.2.7}$$

where $\phi \in C_c^\infty(\mathbf{R})$. It is not difficult to show that the limit exists, and is a distribution (Exercise 1.3). Alternatively, one can deduce this from the identity

$$\partial(\log|x|) = x^{-1} \tag{2.2.8}$$

which is proved as follows:

$$\langle \partial(\log|x|), \phi \rangle = - \quad \partial\phi(x) \log|x|\,\mathrm{d}x$$

$$= - \lim_{\epsilon\to 0+} \left(\int_{-\infty}^{-\epsilon} \partial\phi(x) \log(-x)\,\mathrm{d}x \right.$$

$$\left. + \int_{\epsilon}^{\infty} \partial\phi(x) \log x\,\mathrm{d}x \right)$$

$$= \lim_{\epsilon\to 0+} \left((\phi(\epsilon) - \phi(-\epsilon)) \log\epsilon + \int_{-\infty}^{-\epsilon} \frac{\phi(x)}{x}\,\mathrm{d}x \right.$$

$$\left. + \int_{\epsilon}^{\infty} \frac{\phi(x)}{x}\,\mathrm{d}x \right)$$

Here, $\phi \in C_c^\infty(\mathbf{R})$ so that

$$(\phi(\epsilon) - \phi(-\epsilon)) \log\epsilon = O(\epsilon \log\epsilon) \to 0 \text{ as } \epsilon \to 0+$$

As a first application of Theorem 2.1.1, we shall compute the boundary value of z^{-1}, where $z = x + \mathrm{i}y$ is a complex variable and $y > 0$. Define $\log z$ by

$$\log z = \log|z| + \mathrm{i} \arg z, \quad 0 < \arg z < \pi.$$

One can now regard $x \mapsto \log z$ as a distribution which depends on the parameter $y \in \mathbf{R}^+$. For $x \neq 0$ one has the pointwise limit

$$\lim_{y \to 0+} \log z = \log |x| + i\pi(1 - H(x)).$$

By Theorem 1.5.1, this also holds in $\mathscr{D}'(\mathbf{R})$; the proof is left to the reader as an exercise. By Theorem 2.1.1, one can differentiate (with respect to x) and then go to the limit. By (2.2.2) and (2.2.8), this gives

$$\frac{1}{x + i0} = \lim_{y \to 0+} \frac{1}{x + iy} = \frac{1}{x} - i\pi\delta(x), \tag{2.2.9}$$

where $1/x$ stands for the principal value distribution (2.2.7). Similarly

$$\frac{1}{x - i0} = \frac{1}{x} + i\pi\delta(x). \tag{2.2.10}$$

In Euclidean space of dimension $n > 1$ the analogue of (2.2.2) is, in effect, the divergence theorem. Let $\Omega \subset \mathbf{R}^n$ be an open set with a C^1 boundary. This means that for every point $y \in \partial\Omega$ there is a neighbourhood U and an $S \in C^1(U)$ such that $dS \neq 0$ on $\partial\Omega$ and $\Omega \cap U = \{x \in U: S(x) < 0\}$. The divergence theorem gives

$$\int_\Omega \partial_i\phi \, dx = \int_{\partial\Omega} \phi\theta_i \, d\sigma, \quad i = 1, \ldots, n, \quad \phi \in C_c^\infty(\mathbf{R}^n),$$

where $\theta \in S^{n-1}$ is the unit vector normal to $\partial\Omega$ at x oriented such that $\theta \cdot dS > 0$, and σ is the usual ('Euclidean') surface measure on $\partial\Omega$. One can write this as

$$\partial_i\chi = -\theta_i\sigma, \quad i = 1, \ldots, n, \tag{2.2.11}$$

where χ is the characteristic function of Ω,

$$\chi(x) = 1 \text{ if } x \in \Omega, \quad \chi(x) = 0 \text{ if } x \notin \Omega. \tag{2.2.12}$$

Note that the support of σ is $\partial\Omega$. Alternatively, (2.2.11) can be taken as the definition of σ.

2.3. A distribution obtained by analytic continuation

Take $X = \mathbf{R}$ and define a function $x \mapsto x_+$ by setting

$$x_+ = x \text{ if } x > 0, \quad x_+ = 0 \text{ if } x \leqslant 0. \tag{2.3.1}$$

If $\lambda \in \mathbf{C}$ and $\operatorname{Re} \lambda > 0$, then $x \mapsto x_+^{\lambda-1}$ is locally integrable and so determines a distribution,

$$\langle x_+^{\lambda-1}, \phi \rangle = \int_0^\infty x^{\lambda-1}\phi(x) \, dx, \quad \phi \in C_c^\infty(\mathbf{R}). \tag{2.3.2}$$

Here one can differentiate with respect to λ under the integral sign. This actually follows from a theorem on Lebesgue integrals (Theorem 8.2.1 below), but is elementary when Re $\lambda > 1$, as $x_+^{\lambda-1}$ and $x_+^{\lambda-1}\log x$ are then continuous functions. So the second member of (2.3.2) is C^1 and satisfies the Cauchy–Riemann equations, hence is analytic on $\mathbf{C}^+ = \{\lambda \in \mathbf{C}\colon \operatorname{Re}\lambda > 0\}$. Thus (2.3.2) gives an analytic function $\mathbf{C}^+ \to \mathscr{D}'(\mathbf{R})$.

Evidently, one has

$$\partial x_+^{\lambda} = \lambda x_+^{\lambda-1} \tag{2.3.3}$$

when Re $\lambda > 0$. Let us now define a *distribution* $x_+^{\lambda-1}$ by setting

$$x_+^{\lambda-1} = \frac{1}{\lambda(\lambda+1)\ldots(\lambda+k-1)}\,\partial^k x_+^{\lambda+k-1}. \tag{2.3.4}$$

where k is a nonnegative integer chosen so that Re $\lambda + k > 0$. Explicitly, one has

$$\langle x_+^{\lambda-1}, \phi\rangle = \frac{(-1)^k}{\lambda(\lambda+1)\ldots(\lambda+k-1)}\int_0^\infty x^{\lambda+k-1}\partial^k\phi(x)\,dx,$$

$$\phi \in C_c^\infty(\mathbf{R}). \tag{2.3.5}$$

It is clear, by partial integration, that this agrees with (2.3.2) when Re $\lambda > 0$, is independent of the choice of k subject to Re $\lambda + k > 0$, and gives an analytic function of λ on the set

$$\Omega = \{\lambda \in \mathbf{C}\colon \lambda \neq 0, -1, \ldots\} = \mathbf{C}\backslash\{0, -1, \ldots\}, \tag{2.3.6}$$

valued in $\mathscr{D}'(\mathbf{R})$. As this is a connected set, (2.3.4) is the unique analytic continuation of the distribution (2.3.2), considered as an analytic function on $\mathbf{C}^+$, to Ω.

The excluded points $\lambda = 0, -1, \ldots$ are simple poles of this function. To compute the residues at these poles, replace k by $k+1$ in (2.3.5), to obtain

$$\operatorname*{res}_{\lambda=-k}\langle x_+^{\lambda-1}, \phi\rangle = \lim_{\lambda\to -k}(\lambda+k)\langle x_+^{\lambda-1}, \phi\rangle$$

$$= \frac{(-1)^{k+1}}{(-k)(-k+1)\ldots(-1)}\int_0^\infty \partial^{k+1}\phi(x)\,dx = \partial^k\phi(0)/k!,$$

by dominated convergence. Hence

$$\operatorname*{res}_{\lambda=-k} x_+^{\lambda-1} = (-1)^k\partial^k\delta/k!. \tag{2.3.7}$$

We also note that (2.3.3) holds in $\mathscr{D}'(\mathbf{R})$ for all $\lambda \in \Omega$. This gives an alternative proof of (2.3.7), since $x_+^0 = H$, the Heaviside function.

The gamma function $\Gamma(\lambda)$ has the same poles as the distribution $x_+^{\lambda-1}$, and its residue at the point $\lambda = -k$ is $(-1)^k/k!$. Define $E_\lambda \in \mathscr{D}'(\mathbf{R})$ by

$$E_\lambda = x_+^{\lambda-1}/\Gamma(\lambda), \quad \lambda \in \Omega; \quad E_{-k} = \partial^k\delta, \quad k = 0, 1, \ldots. \tag{2.3.8}$$

Then $\lambda \to E_\lambda$ is analytic on Ω and continuous on $\mathbf{C}$. An elementary argument shows that E_λ is therefore analytic on $\mathbf{C}$. Note that one has, by (2.3.3) and the functional equation of the gamma function,

$$\partial E_\lambda = E_{\lambda-1}, \quad \lambda \in \mathbf{C}. \tag{2.3.9}$$

2.4. Primitives in $\mathscr{D}'(\mathbf{R})$

Let us again take $X = \mathbf{R}$. The simplest differential equation is

$$\partial u = v, \tag{2.4.1}$$

where $v \in \mathscr{D}'(\mathbf{R})$ is given. A solution of this equation is called a *primitive* of v.

Now (2.4.1) means that one is seeking $u \in \mathscr{D}'(\mathbf{R})$ such that

$$\langle u, \partial\phi\rangle = -\langle v, \phi\rangle, \quad \phi \in C_c^\infty(\mathbf{R}).$$

This determines u on

$$\partial C_c^\infty(\mathbf{R}) = \{\psi \in C_c^\infty(\mathbf{R}) \text{: there is a } \phi \in C_c^\infty(\mathbf{R}) \text{ such that } \psi = \partial\phi\}, \tag{2.4.2}$$

which is a subspace of $C_c^\infty(\mathbf{R})$. To solve (2.4.1) one must therefore extend the linear form $\psi \mapsto -\langle v, \phi\rangle$ from $\partial C_c^\infty(\mathbf{R})$ to $C_c^\infty(\mathbf{R})$. Now

$$\partial C_c^\infty(\mathbf{R}) = \{\psi \in C_c^\infty(\mathbf{R}) \text{: } \langle 1, \psi\rangle = \int \psi \,\mathrm{d}x = 0\} \tag{2.4.3}$$

The implication (2.4.2) $\Rightarrow$ (2.4.3) is trivial; conversely, if $\psi \in C_c^\infty(\mathbf{R})$ and $\langle 1, \psi\rangle = 0$, then

$$\phi(x) = \int_{-\infty}^{x} \psi(t)\,\mathrm{d}t = -\int_{x}^{\infty} \psi(t)\,\mathrm{d}t$$

is in $C^\infty(\mathbf{R})$ and has compact support, contained in the convex hull of supp ψ, and $\partial\phi = \psi$.

Chose some $\phi_0 \in C_c^\infty(\mathbf{R})$ such that $\langle 1, \phi_0\rangle = 1$, and decompose any $\phi \in C_c^\infty(\mathbf{R})$ as follows:

$$\phi(x) = (\phi(x) - \langle 1, \phi\rangle\phi_0(x)) + \langle 1, \phi\rangle\phi_0(x). \tag{2.4.4}$$

This amounts to representing $C_c^\infty(\mathbf{R})$ as the direct sum of $\partial C_c^\infty(\mathbf{R})$ and the one-dimensional subspace generated by ϕ_0. One can then define a map μ: $C_c^\infty(\mathbf{R}) \to C_c^\infty(\mathbf{R})$ by setting

$$\mu\phi(x) = \int_{x}^{\infty} (\phi(t) - \langle 1, \phi\rangle\phi_0(t))\,\mathrm{d}t. \tag{2.4.5}$$

It is clear from (2.4.4) and (2.4.5) that, if (2.4.1) has a solution u, then

$$\langle u, \phi\rangle = \langle v, \mu\phi\rangle + \langle C, \phi\rangle, \quad C = \langle u, \phi_0\rangle. \tag{2.4.6}$$

Theorem 2.4.1. If $v \in \mathscr{D}'(\mathbf{R})$, then u, defined by (2.4.6), with C an arbitrary complex number, is a primitive of v; moreover, every primitive of v is of this form.

Proof. We first show that u is a distribution. To prove this it is sufficient to establish a semi-norm estimate in any closed (bounded) interval $[-a, a]$. Now if $\phi \in C_c^\infty[-a, a]$ then, obviously,

$$\operatorname{supp} \mu\phi \subset [-a, a] \cup \operatorname{supp} \phi_0,$$

$$\sup |\mu\phi| \leqslant 2a(1 + \sup |\phi_0|) \sup |\phi|,$$

$$\sup |\partial^k \mu\phi| \leqslant \sup |\partial^{k-1}\phi| + 2a \sup |\partial^{k-1}\phi_0| \sup |\phi|, \quad k \geqslant 1.$$

So a semi-norm estimate for u on $[-a, a]$ can be derived from one for v on $[-a, a] \cup \operatorname{supp} \phi_0$, whence $u \in \mathscr{D}'(\mathbf{R})$. Also, (2.4.5) gives $\mu\partial\phi = -\phi$, so that our u satisfies (2.4.1).

Finally, if u_1 and u_2 are both primitives of v then $u = u_1 - u_2$ satisfies $\partial u = 0$, and as the first term on the right hand side of (2.4.4) is a member of $\partial C_c^\infty(\mathbf{R})$, it follows that $u = C$, where $C = \langle u, \phi_0 \rangle$ is a constant which depends on the choice of ϕ_0. So the theorem is proved.

The argument also works in any open interval, and hence in any open set $X \subset \mathbf{R}$. But then the respective constants C in the intervals of which X consists are not related.

2.5. Product of a distribution and a smooth function

We now return to the general setting where $X \subset \mathbf{R}^n$ is an open set. If u and f are continuous functions on X, then so is fu, and formally

$$\langle fu, \phi \rangle = \int fu\phi \, dx = \langle u, f\phi \rangle.$$

In order to extend this to distributions, one has to take the multiplier f to be C^∞. (There is no general definition of the product of two distributions.) One then has $f\phi \in C_c^\infty(X)$ if $\phi \in C_c^\infty(X)$, as $\operatorname{supp} f\phi \subset \operatorname{supp} \phi$, and $f\phi \in C^\infty(X)$ by Leibniz's theorem. This theorem also implies that, if $K \subset X$ is a compact set and $\phi \in C_c^\infty(K)$, then there are constants $C_0, C_1, \ldots$, depending on f and K but not on ϕ, such that, for $N = 0, 1, \ldots$

$$\sum_{|\alpha| \leqslant N} \sup |\partial^\alpha(f\phi)| \leqslant C_N \sum_{|\alpha| \leqslant N} \sup |\partial^\alpha \phi|, \quad \phi \in C_c^\infty(K). \tag{2.5.1}$$

Hence $\phi \mapsto \langle u, f\phi \rangle$ is a distribution if u is one, and we have justified the following definition:

Definition 2.5.1. If $u \in \mathscr{D}'(X)$ and $f \in C^\infty(X)$, then the distribution

$$\langle fu, \phi \rangle = \langle u, f\phi \rangle, \quad \phi \in C_c^\infty(X), \tag{2.5.2}$$

is called the product of f and u.

It is an immediate consequence of this definition and Theorem 1.3.1 that the distribution fu is equal to the pointwise product fu when $u \in C^0(X)$. We also have the following, from (2.5.1) and (2.5.2):

Theorem 2.5.1. If $u \in \mathscr{D}'(X)$ and $f \in C^\infty(X)$, then the map $u \mapsto fu$ is a sequentially continuous map $\mathscr{D}'(X) \to \mathscr{D}'(X)$.

Leibniz's theorem extends to the product which has just been defined, and can be given a concise form in the multi-index notation.

Theorem 2.5.2. Let $u \in \mathscr{D}'(X), f \in C^\infty(X)$, and let α be a multi-index. Then

$$\partial^\alpha(fu) = \sum_{\beta+\gamma=\alpha} \frac{\alpha!}{\beta!\gamma!} \partial^\beta f \partial^\gamma u. \tag{2.5.3}$$

Proof. This identity is trivial for $\alpha = 0$. When $|\alpha| = 1$, it is obtained by a simple manipulation. Indeed,

$$\langle \partial_i(fu), \phi \rangle = -\langle fu, \partial_i \phi \rangle = -\langle u, f\partial_i \phi \rangle, \quad i = 1, \ldots, n, \quad \phi \in C_c^\infty(X).$$

But

$$-f\partial_i \phi = -\partial_i(f\phi) + \phi \partial_i f$$

so

$$\langle \partial_i(fu), \phi \rangle = \langle \partial_i u, f\phi \rangle + \langle u, \phi \partial_i f \rangle$$

whence

$$\partial_i(fu) = f\partial_i u + (\partial_i f), \quad i = 1, \ldots, n. \tag{2.5.4}$$

From this it follows by induction that

$$\partial^\alpha(fu) = \sum_{\beta+\gamma=\alpha} C_{\beta\gamma}^\alpha \partial^\beta f \partial^\gamma u$$

where the $C_{\beta\gamma}^\alpha$ are positive integers. To determine these, take $f = \exp(x \cdot \xi)$ and $u = \exp(x \cdot \eta)$, where $\xi \in \mathbf{R}^n$ and $\eta \in \mathbf{R}^n$. Cancelling the common factor $\exp(x \cdot (\xi + \eta))$, one obtains

$$(\xi + \eta)^\alpha = \sum C_{\beta\gamma}^\alpha \xi^\beta \eta^\gamma$$

and as the monomials $\xi^\beta \eta^\gamma$ are linearly independent, (2.5.3) now follows from Taylor's theorem for polynomials.

As a simple example, let us take $X = \mathbf{R}^n$ and compute $f\delta$:

$$\langle f\delta, \phi \rangle = \langle \delta, f\phi \rangle = f(0)\phi(0), \quad \phi \in C_c^\infty(\mathbf{R}^n).$$

So

$$f\delta = f(0)\delta. \tag{2.5.5}$$

Again, (2.5.4) gives

$$\partial_i(f\delta) = (\partial_i f)\delta + f\partial_i\delta, \quad i = 1, \ldots, n,$$

whence, by (2.5.5),

$$f\partial_i\delta = f(0)\partial_i\delta - \partial_i f(0)\delta, \quad i = 1, \ldots, n. \tag{2.5.6}$$

2.6. Linear differential operators

Let A be a finite set of multi-indices, and let $(a_\alpha(x))_{\alpha\in A} \in C^\infty(X)$ be functions indexed by α. Then

$$P(x, \partial) = \sum_{\alpha\in A} a_\alpha(x)\partial^\alpha$$

is a *linear differential operator* with C^∞ coefficients, defined on X, 'ordinary' or 'partial' according as $n = 1$ or $n > 1$. The *order* of P is max $\{|\alpha|: \alpha\in A\}$. We denote this by m and write

$$P(x, \partial) = \sum_{|\alpha|\leqslant m} a_\alpha(x)\partial^\alpha, \tag{2.6.1}$$

with the understanding that, while some of the a_α may vanish identically, there is at least one $a_\alpha \equiv 0$ with $|\alpha| = m$.

One should think of a differential operator as a map. In particular, it follows from Theorems 2.1.1 and 2.5.1 that $u \mapsto Pu$ is a sequentially continuous map $\mathscr{D}'(X) \to \mathscr{D}'(X)$. Explicitly, one has

$$\langle Pu, \phi\rangle = \langle u, {}^tP\phi\rangle, \quad \phi \in C_c^\infty(X), \tag{2.6.2}$$

where ${}^tP$ is the differential operator

$${}^tP\phi = \sum_{|\alpha|\leqslant m} (-1)^\alpha \partial^\alpha(a_\alpha\phi). \tag{2.6.3}$$

It is called the *adjoint* of P.

When $u \in C_c^\infty(X)$, then the identity (2.6.2) becomes

$$\int \phi Pu \,\mathrm{d}x = \int u\,{}^tP\phi \,\mathrm{d}x$$

By iteration one obtains

$$\int (Pu - {}^t({}^tP)u)\phi \,\mathrm{d}x = 0, \quad u, \phi \in C_c^\infty(X).$$

Hence $Pu - {}^t({}^tP)u = 0$ if $u \in C_c^\infty(X)$. Let y be a point of X, and let $\psi \in C_c^\infty(X)$

be such that $\psi = 1$ on a neighbourhood of y. Taking $u = \psi(x)(x-y)^\alpha/\alpha!$ and setting $x = y$ one concludes that the coefficients of ∂^α in P and ${}^t({}^tP)$ are the same at y, and as y can be any point of X it follows that

$$ {}^t({}^tP) = P. \tag{2.6.4}$$

Leibniz's theorem can be extended to differential operators, as follows. For an operator P given by (2.6.1), the function

$$P(x,\xi) = \sum_{|\alpha|\leq m} a_\alpha(x)\xi^\alpha : X \times \mathbf{R}^n \to \mathbf{C} \tag{2.6.5}$$

is called the *symbol* of P. Now (2.5.3) gives

$$P(x,\partial)(fu) = \sum_{|\alpha|\leq m} a_\alpha(x) \sum_{\beta+\gamma=\alpha} \frac{\alpha!}{\beta!\gamma!} \partial^\beta f \partial^\gamma u.$$

After some rearrangement, this can be put into the form

$$P(x,\partial)(fu) = \sum_{\alpha\geq 0} \frac{\partial^\alpha f}{\alpha!} P^{(\alpha)}(x,\partial)u = \sum_{\alpha\geq 0} \frac{\partial^\alpha u}{\alpha!} P^{(\alpha)}(x,\partial)f \tag{2.6.6}$$

where $P^{(\alpha)}(x,\partial)$ is the differential operator whose symbol is

$$P^{(\alpha)}(x,\xi) = \partial_\xi^\alpha P(x,\xi). \tag{2.6.7}$$

The symbol (2.6.5) is sometimes called the full symbol of P, to distinguish it from the *principal symbol*

$$\sigma_P(x,\xi) = \sum_{|\alpha|=m} a_\alpha(x)\xi^\alpha. \tag{2.6.8}$$

There is a straightforward extension to systems of differential operators. Let $N > 1$ be an integer, and let $(\mathscr{D}'(X))^N$ be the direct sum of N copies of $\mathscr{D}'(X)$, so that one has a map

$$\phi \mapsto \langle u,\phi\rangle = (\langle u_1,\phi\rangle, \ldots, \langle u_N,\phi\rangle), \quad \phi \in C_c^\infty(X) \tag{2.6.9}$$

in which each u_j is a distribution. Now if $P = (P_{jk})_{1\leq j,k\leq N}$ is a matrix whose entries are differential operators (with C^∞ coefficients) one has a map $(\mathscr{D}'(X))^N \to (\mathscr{D}'(X))^N$ given by

$$(u_1,\ldots,u_N) \mapsto \left(\sum_{k=1}^N P_{jk}(x,\partial)u_k\right)_{j=1,\ldots,N}, \tag{2.6.10}$$

which one can either think of as a system of differential operators, or as a single operator on $(\mathscr{D}'(X))^N$. To write down the adjoint, it is best to introduce the space $(C_c^\infty(X))^N$ and the inner product

$$\langle u,\phi\rangle = \sum_{j=1}^N \langle u_j,\phi_j\rangle, \quad u \in (\mathscr{D}'(X))^N, \quad \phi \in (C_c^\infty(X))^N.$$

Then

$$\langle Pu, \phi\rangle = \langle u, {}^tP\phi\rangle \tag{2.6.11}$$

where ${}^tP$ is the transpose of the matrix whose entries are the adjoints of the entries of P. Under- and overdetermined systems can be handled in the same way.

2.7. Division in $\mathscr{D}'(\mathbf{R})$

The *division problem* is this: given $v \in \mathscr{D}'(X)$ and $f \in C^\infty(X)$, to find $u \in \mathscr{D}'(X)$ such that $fu = v$. If $f \neq 0$ on X, then $x \mapsto 1/f$ is C^∞ and $1/|f|$ is bounded away from zero on every compact subset of X, and one just has $u = v/f$. But if $f^{-1}\{0\} = \{x: f(x) = 0\}$ is not empty, and $n > 1$, then the division problem is, in general, not an elementary one. However, it is not difficult to solve it when $X = \mathbf{R}$ and $f^{-1}\{0\}$ consists of isolated zeros of finite order. By a partition of unity and a translation of the origin this reduces in effect to the case where $X = \mathbf{R}$ and $f = x^m$, where m is a positive integer. One then has the following result.

Theorem 2.7.1. Let m be a positive integer. (i) If $u \in \mathscr{D}'(\mathbf{R})$ and $x^m u = 0$, then

$$u = \sum_{j=0}^{m-1} c_j \partial^j \delta, \tag{2.7.1}$$

where the c_j are complex numbers.

(ii) If $v \in \mathscr{D}'(\mathbf{R})$ is given, then one can find $u \in \mathscr{D}'(\mathbf{R})$ such that

$$x^m u = v \tag{2.7.2}$$

Note. Combining (i) and (ii), one sees that, if w is any particular solution of (2.7.2), then the general solution of this equation is

$$u = w + \sum_{j=0}^{m-1} c_j \partial^j \delta$$

where the c_j are 'arbitrary constants'.

Proof. The equation (2.7.2) means that

$$\langle u, x^m\phi\rangle = \langle v, \phi\rangle, \quad \phi \in C_c^\infty(\mathbf{R}).$$

This certainly defines a linear form on

$$\{\phi \in C_c^\infty(\mathbf{R}): \quad \phi = x^m\psi, \quad \psi \in C_c^\infty(\mathbf{R})\}$$

which is a subspace of $C_c^\infty(\mathbf{R})$, and – as in the problem of finding the primitives of a distribution – the question is one of extending this linear form to $C_c^\infty(\mathbf{R})$. For this, one needs a lemma.

Lemma 2.7.1. Suppose that $\phi \in C_c^\infty(\mathbf{R})$ and that $\partial^j\phi(0) = 0$ for $j = 1, \ldots, m-1$.

Then

$$\phi = x^m \psi, \quad \psi \in C_c^\infty(\mathbf{R}) \tag{2.7.3}$$

and one has

$$\sup |\partial^k \psi| \leqslant \frac{1}{(m-1)!} \sup |\partial^{m+k}\phi|, \quad k=0,1,\ldots \tag{2.7.4}$$

Proof of the lemma. One starts with Taylor's formula,

$$\phi(x) = \sum_{j=0}^{m-1} \frac{x^j}{j!} \partial^j \phi(0) + \frac{1}{(m-1)!}\int_0^x (x-t)^{m-1}\partial^m \phi(t)\, dt, \tag{2.7.5}$$

which is proved by integration by parts. By hypothesis, $\partial^j\phi(0)=0$ for $j=0,1,\ldots,m-1$, so the change of variable of integration $t \mapsto xs$ gives (2.7.3) with

$$\psi(x) = \frac{1}{(m-1)!}\int_0^1 (1-s)^{m-1}\phi^{(m)}(xs)\, ds, \tag{2.7.6}$$

where $\phi^{(m)}$ stands for $\partial^m \phi$. It is obvious that $\psi \in C^\infty(\mathbf{R})$, as one can differentiate under the integral sign. The estimates (2.7.4) also follow from (2.7.6), since $0 \leqslant s^k(1-s)^{m-1} \leqslant 1$ when $0 \leqslant s \leqslant 1$. Finally, it is clear from $\phi = x^m\psi$ that ϕ and ψ restricted to $\mathbf{R}\setminus\{0\}$ have the same support, hence supp $\psi =$ supp ϕ whence $\psi \in C_c^\infty(\mathbf{R})$. So the lemma is proved.

Proof of the theorem. Chose $\phi_0 \in C_c^\infty(\mathbf{R})$ such that $\phi_0 = 1$ on a neighbourhood of $x=0$. Any $\phi \in C_c^\infty(\mathbf{R})$ can then be written as

$$\phi(x) = \phi_0(x)\sum_{j=0}^{m-1}\frac{x^j}{j!}\partial^j\phi(0) + \chi(x),$$

say, where χ satisfies the hypotheses of the lemma. Hence one has a map $\mu\colon C_c^\infty(\mathbf{R}) \to C_c^\infty(\mathbf{R})$ such that

$$\phi(x) = \phi_0(x)\sum_{j=0}^{m-1}\frac{x^j}{j!}\partial^j\phi(0) + x^m\mu\phi(x). \tag{2.7.7}$$

If $u \in \mathscr{D}'(\mathbf{R})$ and $x^m u = 0$, then this gives

$$\langle u, \phi\rangle = \sum_{j=0}^{m-1}\frac{1}{j!}\langle u, x^j\phi_0\rangle \partial^j\phi(0),$$

which is (2.7.1), with certain constants c_j that depend on the choice of ϕ_0. On the other hand, it is an easy exercise to show that $x^m\partial^j\delta = 0$ for $j<m$. This settles part (i).

To prove part (ii), define u by

$$\langle u, \phi\rangle = \langle v, \mu\phi\rangle, \quad \phi \in C_c^\infty(\mathbf{R}). \tag{2.7.8}$$

It follows from Lemma 2.7.1 and (2.7.7) that $u\in\mathscr{D}'(\mathbf{R})$; the proof is left as an exercise. It is also clear from (2.7.7) that $\mu(x^m\phi)=\phi$ if $\phi\in C_c^\infty(\mathbf{R})$, so that (2.7.8) solves (2.7.2), and so we are done.

2.8. Duality

In this chapter, two principles have been used repeatedly which are of some importance in the theory of distributions. The first of these is that the transpose of a continuous map between test functions is a sequentially continuous map between distributions. We need a definition for this.

Definition 2.8.1. Let $X\subset\mathbf{R}^n$ and $Y\subset\mathbf{R}^n$ be open sets. A homomorphism μ: $C_c^\infty(Y)\rightarrow C_c^\infty(X)$ is called continuous if (i) for every compact set $K\subset Y$ there is a compact set $K'\subset X$ such that supp $\mu\phi\subset K'$ if supp $\phi\subset K$, and (ii) for any compact set $K\subset Y$ and any nonnegative integer N there is a $C=C(K,N)$ and an integer $M=M(K,N)\geqslant 0$ such that

$$\sum_{|\alpha|\leqslant N}\sup|\partial^\alpha(\mu\phi)|\leqslant C\sum_{|\beta|\leqslant M}\sup|\partial^\beta\phi|,\quad \phi\in C_c^\infty(K). \tag{2.8.1}$$

By imitating the proof of Theorem 1.3.2, it is not hard to show that μ is continuous if and only if it is sequentially continuous, that is to say, maps sequences converging to 0 in $C_c^\infty(Y)$ to sequences converging to 0 in $C_c^\infty(X)$.

Given such a map μ, one can define the transpose ${}^t\mu$: $\mathscr{D}'(X)\rightarrow\mathscr{D}'(Y)$ in the usual way, by setting

$$\langle {}^t\mu u,\phi\rangle=\langle u,\mu\phi\rangle,\quad u\in\mathscr{D}'(X),\quad \phi\in C_c^\infty(Y). \tag{2.8.2}$$

This is well defined, since $\mu\phi\in C_c^\infty(X)$.

Theorem 2.8.1. Let X, Y and μ be as in Definition 2.8.1, and let ${}^t\mu$ be given by (2.8.2). Then ${}^t\mu$ is a sequentially continuous map $\mathscr{D}'(X)\rightarrow\mathscr{D}'(Y)$.

Proof. If $\phi\in C_c^\infty(K)$, where $K\subset Y$ is a compact set, then $\mu\phi\in C_c^\infty(K')$ by (i) of the definition. Hence one has a semi-norm estimate

$$|\langle u,\mu\phi\rangle|\leqslant C'\sum_{|\alpha|\leqslant N}\sup|\partial^\alpha(\mu\phi)|$$

for some $C',N\geqslant 0$. But then (2.8.1) gives a semi-norm estimate for ${}^t\mu u$, and hence ${}^t\mu u\in\mathscr{D}'(Y)$. That ${}^t\mu u_k\rightarrow{}^t\mu u$ in $\mathscr{D}'(Y)$ if $u_k\rightarrow u$ in $\mathscr{D}'(X)$ as $k\rightarrow\infty$ is obvious; so we are done.

In sections 2.4 and 2.7 this result has been used as it stands. In the definition of differentiation, and of multiplication by a C^∞ function, it has been combined with another observation, which can be formulated concisely in the following way. Let $P=P(x,\partial)$ be a differential operator with smooth coefficients, defined on a set $X \subset \mathbf{R}^n$. It is easy to verify that $P\colon C_c^\infty(X) \to C_c^\infty(X)$ is a continuous map, in the sense of Definition 2.8.1. Now one can apply this to the adjoint differential operator ${}^tP$, and *define* $P\colon \mathscr{D}'(X) \to \mathscr{D}'(X)$ as the transpose map of ${}^tP$; this, in fact, is (2.6.2). It is clear that this definition falls within the scope of Theorem 2.8.1. Taking $P = \partial_i$, $i = 1, \ldots, n$, one recovers Definition 2.1.1, and taking $P = f \in C^\infty(X)$ one gets Definition 2.5.1. The principle involved here is that of extending an operator on C_c^∞ to an operator on $\mathscr{D}'$ *by duality*.

Exercises

2.1. Show that

$$\frac{1}{\pi}\frac{\epsilon}{x^2+\epsilon^2} \to \delta \quad \text{in } \mathscr{D}'(\mathbf{R}) \text{ as } \epsilon \to 0+.$$

2.2. Show that

$$f_t(x) = \frac{\sin xt}{\pi x} \to \delta \quad \text{in } \mathscr{D}'(\mathbf{R}) \text{ as } t \to \infty.$$

Deduce that, if $\phi \in C_c^\infty(\mathbf{R})$, then

$$\lim_{t\to\infty} \int_{-t}^{t} \mathrm{d}\xi \int \phi(x) \mathrm{e}^{-\mathrm{i}x\xi}\, \mathrm{d}x = 2\pi\phi(0).$$

$$\left(\text{Suggestion: consider } F_t(x) = \int_{-1}^{x} f_t(s)\, \mathrm{d}s.\right)$$

2.3. Show that, if $u \in \mathscr{D}'(\mathbf{R})$ and $x\partial u + u = 0$, then

$$u = A(1/x) + B\delta$$

where A and B are complex numbers, and $1/x$ is the principal value distribution.

2.4. Show that

$$\left\langle \frac{1}{x}, \phi \right\rangle = \int_0^\infty \frac{\phi(x) - \phi(-x)}{x}\, \mathrm{d}x, \quad \phi \in C_c^\infty(\mathbf{R})$$

and

$$\left\langle \frac{1}{x^2}, \phi \right\rangle = \int_0^\infty \frac{\phi(x) + \phi(-x) - 2\phi(0)}{x^2}\, \mathrm{d}x, \quad \phi \in C_c^\infty(\mathbf{R}),$$

where $1/x^2 = -\partial(1/x)$, and the integrands are defined by continuity at $x = 0$. Derive similar representations of $\langle x^{-m-1}, \phi \rangle$, $\phi \in C_c^\infty(\mathbf{R})$, where $x^{-m-1} = (-1)^m \partial^m (x^{-1})/m!$. (Here $x^{-1} = 1/x$ is the principal value distribution.)

2.5. Let $k \in C^0(\mathbf{R}^n \backslash \{0\})$ be a function such that

$$k(xt) = t^{-n} k(x) \text{ for } 0 \neq x \in \mathbf{R}^n, \quad t > 0.$$

Show that the principal value

$$\text{p.v.} \int k(x)\phi(x)\, \mathrm{d}x = \lim_{\epsilon\to 0+} \int_{|x|>\epsilon} k(x)\phi(x)\, \mathrm{d}x, \quad \phi \in C_c^\infty(\mathbf{R}^n), \tag{$*$}$$

exists if and only if

$$\int_{S^{n-1}} k(\theta)\, d\omega(\theta) = 0, \tag{**}$$

where $d\omega$ is the usual surface measure on the unit sphere S^{n-1}. Show also that (*) is a distribution when (**) holds.

Let $v_1, \ldots, v_n$ be functions of class $C^1(\mathbf{R}^n \backslash \{0\})$ such that

$v_i(xt) = t^{1-n} v_i(x)$ for $0 \neq x \in \mathbf{R}^n, \quad t > 0, \quad i = 1, \ldots, n.$

Show that, for $\phi \in C_c^\infty(\mathbf{R}^n)$,

$$\sum_{i=1}^n \langle \partial_i v, \phi \rangle = \text{p.v.} \int \phi \sum_{i=1}^n \partial_i v_i \, dx + \phi(0) \int_{S^{n-1}} \sum_{i=1}^n \theta_i v_i(\theta)\, d\omega(\theta).$$

2.6. Let $I \subset \mathbf{R}$ be an open interval. Show that, if $u \in \mathscr{D}'(I)$ and $\partial u \in C^\infty(I)$, then $u \in C^\infty(I)$.

Let

$P(x, \partial) = a_0(x)\partial^m + a_n(x)\partial^{m-1} + \cdots + a_m(x)$

be a differential operator with C^∞ coefficients, defined on I, and suppose that $a_0 \neq 0$ on I. Show that, if $u \in \mathscr{D}'(I)$ and $Pu \in C^\infty(I)$, then $u \in C^\infty(I)$.

(Assume that, for each $x_0 \in I$, there is a $\phi \in C^\infty(I)$ such that $P\phi = 0$ and $\phi(x_0) = 1$, and use induction. The existence of such ϕ follows from the theory of ordinary differential equations.)

2.7. Let $x \in \mathbf{R}^n$, and let α, β be multi-indices. Show that

$$\partial^\beta(x^\alpha/\alpha!) = x^{\alpha-\beta}/(\alpha - \beta)! \quad \text{if } \beta \leqslant \alpha$$
$$= 0 \quad \text{if } \beta > \alpha.$$

2.8. Let P be a linear differential operator with C^∞ coefficients, defined on $\mathbf{R}^n$. Suppose that $P\delta = 0$. Show that P can be put into the form

$$P = \sum_{j=1}^n P_j x_j \quad \text{i.e. } Pu = \sum_{j=1}^n P_j(x, \partial)(x_j u),$$

where the $P_j, j = 1, \ldots, n$ are differential operators.

2.9. Show that, if $u \in \mathscr{D}'(\mathbf{R}^n)$ and

$x_1 u = 0, \ldots, x_n u = 0,$

then u is a constant multiple of the Dirac distribution.

2.10. Find all $u \in \mathscr{D}'(\mathbf{R})$ such that $x^m u = \delta$.

2.11. Find all $u \in \mathscr{D}'(\mathbf{R})$ such that $u \sin \pi x = 1$.

2.12. Let λ be a complex number, and k a positive integer such that $\operatorname{Re} \lambda > -k$. Show that (2.3.4) is equivalent to

$$\langle x_+^{\lambda-1}, \phi \rangle = \int_0^1 x^{\lambda-1}\left(\phi(x) - \sum_{j=0}^{k-1} \frac{x^j}{j!} \partial^j \phi(0)\right) dx + \int_1^\infty x^{\lambda-1}\phi(x)\, dx$$
$$+ \sum_{j=0}^{k-1} \frac{\partial^j \phi(0)}{j!(j+\lambda)}, \quad \phi \in C_c^\infty(\mathbf{R}).$$

Hence show that if $-k < \operatorname{Re}\lambda < -k+1$ and $\epsilon > 0$ then

$$\int_\epsilon^\infty x^{\lambda-1}\phi(x)\,dx = \langle x_+^{\lambda-1}, \phi\rangle + \sum_{j=0}^{k-1} \langle u_j, \phi\rangle \epsilon^{\lambda+j} + 0(\epsilon^{\operatorname{Re}\lambda+k-1}),$$

where the $u_j, j=1,\ldots,k-1$ are certain distributions.

2.13. Let $k \geqslant 0$ be an integer, and $\lambda \in \mathbf{C}$. Show that, if $|\lambda+k|<1$, then

$$x_+^{\lambda-1} = \frac{(-1)^k}{k!}\left(\frac{\partial^k\delta}{\lambda+k} + c_k\partial^k\delta + \partial^{k+1}\log_+ x + \rho_\lambda\right)$$

where

$$c_0 = 1, \quad c_k = 1 + \frac{1}{2} + \cdots + \frac{1}{k} \text{ if } k>0,$$

$$\log_+ x = \log x \text{ if } x>0, \quad = 0 \text{ if } x \leqslant 0,$$

and $\rho_\lambda \to 0$ in $\mathscr{D}'(\mathbf{R})$ as $\lambda \to -k$.

2.14. Define the function $x \mapsto x_-$ on $\mathbf{R}$ by $x_- = (-x)_+$. Show that, for $\lambda \in \mathbf{C}\backslash\{0,-1,\ldots\}$ there is a distribution $x_-^{\lambda-1}$ which is an analytic function of λ and equal to the locally integrable function $x_-^{\lambda-1}$ when $\operatorname{Re}\lambda>0$. Show that it has simple poles at $\lambda=0,-1,\ldots$, and calculate the residues at these.

Put, also,

$$|x|^{\lambda-1} = x_+^{\lambda-1} + x_-^{\lambda-1}$$

$$(x)^{\lambda-1}\operatorname{sign} x = x_-^{\lambda-1} - x_-^{\lambda-1}.$$

Determine the regions of $\mathbf{C}$ on which these distributions are defined and analytic in λ, and compute the residues at the poles.

2.15. (i) Determine $(x+i0)^{-m}$ when m is a positive integer. (For $m=1$, see (2.2.9).)

(ii) Define, for $y>0$,

$$(x+iy)^{\lambda-1} = \exp(\lambda-1)\log(x+iy))$$

with $\log(x+i0)$ real when $x>0$. Show that, if $\lambda \neq 0,-1,\ldots$ then, in $\mathscr{D}'(\mathbf{R})$,

$$(x+i0)^{\lambda-1} = \lim_{y\to 0+}(x+iy)^{\lambda-1} = x_+^{\lambda-1} - e^{i\pi\lambda}x^{\lambda-1}.$$

(iii) Deduce that $(x+i0)^{\lambda-1}$ is an analytic function of λ on all of $\mathbf{C}$.

2.16. Let $I \subset \mathbf{R}$ be an open interval, and let $f(z)=f(x+iy)$ be an analytic function, defined on $I \times (0,b)$ for some $b>0$. Show that, if $\phi \in C_c^\infty(I)$ and $0<y_1<y_2<b$, then

$$\int f(x+iy_1)\phi(x)\,dx - \int f(x+iy_2)\phi(x)\,dx = i\int_I\int_{y_1}^{y_2} f(x+iy)\left(\frac{\partial\phi}{\partial x} + i\frac{\partial\phi}{\partial y}\right)dx\,dy.$$

Suppose now that there are real numbers $C \geqslant 0$ and λ such that

$$|f(z)| \leqslant Cy^{-\lambda}, \quad z \in I \times (0,b). \qquad (*)$$

Prove that

$$\lim_{y\to 0+}\int f(x+iy)\phi(x)\,dx = \langle f(x+i0), \phi(x)\rangle, \quad \phi \in C_c^\infty(I)$$

exists, and is a distribution.

(Suggestion: Consider the cases $\lambda < 1$ and $\lambda \geqslant 1$ separately.

For $\lambda \geqslant 1$, obtain an estimate for

$$f_1(z) = \int_{z_0}^{z} f(\zeta)\, d\zeta$$

in $I \times (0, b)$ where $z_0 \in I \times (0, b)$ and the integral is over a suitable path. Use induction, and note that if (*) holds, then also $|f(z)| \leqslant Cy^{-\mu}$ if $\mu > \lambda$ and $0 < y \leqslant 1$.)

3 DISTRIBUTIONS WITH COMPACT SUPPORT

There are a number of important subspaces of the vector space of distributions, such as the distributions of finite order. This section deals with another example, the distributions with compact support. It also includes the theorem that a distribution whose support is a point is a finite linear combination of derivatives of the Dirac distribution, a result with useful applications.

3.1. Continuous linear forms on C^∞ and distributions with compact support

We begin with a definition.

Definition 3.1.1. Let $X \subset \mathbf{R}^n$ be an open set. A linear form u on the vector space $C^\infty(X)$ is called continuous if there are a compact set $K \subset X$, a constant $C \geqslant 0$ and a nonnegative integer N such that, for all $\phi \in C^\infty(X)$

$$|\langle u, \phi\rangle| \leqslant C \sum_{|\alpha| \leqslant N} \sup \{|\partial^\alpha \phi| : x \in K\}. \tag{3.1.1}$$

The vector space of continuous linear forms on $C^\infty(X)$ is called $\mathscr{E}'(X)$.

As in the case of distributions, there is an alternative characterization. For this, one needs another definition.

Definition 3.1.2. A sequence $(\phi_k)_{1 \leqslant k < \infty} \in C^\infty(X)$ is said to converge to zero in $C^\infty(X)$ as $k \to \infty$ if, for each multi-index α, the $\partial^\alpha \phi_k$ converge to zero uniformly on every compact subset of X.

An argument similar to the proof of Theorem 1.3.2, which is left as an exercise, then gives:

Theorem 3.1.1. One has $u \in \mathscr{E}'(X)$ if and only if $\langle u, \phi_k\rangle \to 0$ when $\phi_k \to 0$ in $C^\infty(X)$ as $k \to \infty$.

Remark. The functions

$$\phi \mapsto \|\phi\|_{N,K} = \sum_{|\alpha| \leqslant N} \sup \ |\partial^\alpha \phi| : x \in K \ , \tag{3.1.2}$$

where $N=0, 1, \ldots,$ and K ranges over the compact subsets of X, are semi-norms on $C^\infty(X)$ which generate a topology, and $\mathscr{E}'(X)$ is the dual of $C^\infty(X)$. In fact, the topology is the same as that generated by the $\|\phi\|_{N,K_j}$ when $K_1, K_2, \ldots$ is an increasing sequence of compact sets whose union is X. It is an easy exercise to show that $C^\infty(X)$ is complete. So $C^\infty(X)$, equipped with this topology of uniform convergence of derivatives of all orders on compact sets, is a Fréchet space. One consequence of this is that sequential continuity is equivalent to continuity.

It is obvious that $C_c^\infty(X) \subset C^\infty(X)$, and that the identity map is continuous: a sequence that converges to 0 in $C_c^\infty(X)$ also does so in $C^\infty(X)$. So it follows from Theorems 3.1.1 and 1.3.2 that the restriction of any $u \in \mathscr{E}'(X)$ to $C_c^\infty(X)$ is a distribution. (This also follows easily from Definition 3.1.1.) Now (3.1.1) shows that $\langle u, \phi\rangle = 0$ if the support of ϕ is disjoint from K, so u has compact support, when regarded as a member of $\mathscr{D}'(X)$. In fact, one can prove more:

Theorem 3.1.2. If $u \in \mathscr{D}'(X)$ has compact support, then there is a unique member of $\mathscr{E}'(X)$ whose restriction to $C_c^\infty(X)$ is equal to u.

Proof. Let $\rho \in C_c^\infty(X)$ be chosen such that $\rho = 1$ on a neighbourhood of the support of u. Then one can define a linear form $\tilde{u}$ on $C^\infty(X)$ by setting

$$\langle \tilde{u}, \phi\rangle = \langle u, \rho\phi\rangle, \quad \phi \in C^\infty(X). \tag{3.1.3}$$

To show that this construction gives a unique $\tilde{u}$, one observes that, if $\sigma \in C_c^\infty(X)$ and $\sigma = 1$ on a neighbourhood of the support of u, then

$$\langle u, \rho\phi\rangle - \langle u, \sigma\phi\rangle = \langle u, (\rho - \sigma)\phi\rangle = 0, \quad \phi \in C^\infty(X)$$

by Theorem 1.4.2, since $(\rho - \sigma)\phi = 0$ on a neighbourhood of supp u. The same theorem shows that $\langle \tilde{u}, \phi\rangle = \langle u, \phi\rangle$ when $\phi \in C_c^\infty(X)$. So (3.1.3) extends u to a linear form on $C^\infty(X)$ uniquely.

Furthermore, one has supp $\rho\phi \subset$ supp ρ which is a fixed compact set once ρ has been chosen. So there is a semi-norm estimate

$$|\langle \tilde{u}, \phi\rangle| \leqslant C \sum_{|\alpha| \leqslant N} \sup |\partial^\alpha(\rho\phi)|, \quad \phi \in C^\infty(X).$$

By Leibniz's theorem, this gives a semi-norm estimate (3.1.1) with $K = \operatorname{supp} \rho$ and a different constant C. Hence $\tilde{u} \in \mathscr{E}'(X)$, and the theorem is proved.

Remark. It follows from Theorem 3.1.2 and the remarks preceding it that one can identify $\mathscr{E}'(X)$ with the subspace of $\mathscr{D}'(X)$ consisting of distributions with compact support. It is a standard convention to do so. We shall therefore drop the tilde, and write $\langle u, \phi\rangle$ for the value of $u \in \mathscr{E}'(X)$ at $\phi \in C^\infty(X)$, suppressing the cut-off function ρ. We also note the following obvious consequence:

Corollary. If $u \in \mathscr{E}'(X)$, then u is a distribution of finite order.

There is another observation which is worth recording:

Theorem 3.1.3. $C_c^\infty(X)$ is dense in $C^\infty(X)$, and $\mathscr{E}'(X)$ is dense in $\mathscr{D}'(X)$.

Proof. Let $(\psi_j)_{1 \leqslant j < \infty}$ be a locally finite partition of unity on X. (See the Note at the end of 1.4.) Let $\phi \in C^\infty(X)$ be given and set

$$\phi_k(x) = \sum_{j=1}^{k} \phi(x)\psi_j(x), \quad k = 1, 2, \ldots$$

Then $\phi_k \in C_c^\infty(X)$, and it is an easy exercise to show that $\phi_k \to \phi$ in $C^\infty(X)$ as $k \to \infty$.

Again, if $u \in \mathscr{D}'(X)$ is given, then the distributions

$$u_k = \sum_{j=1}^{k} \psi_j u, \quad k = 1, 2, \ldots$$

are in $\mathscr{E}'(X)$, and obviously converge to u in $\mathscr{D}'(X)$ as $k \to \infty$.

Note. When $X = \mathbf{R}^n$, the proof is simpler still. Choose $\psi \in C_c^\infty(\mathbf{R}^n)$ such that $\psi(x) = 1$ when $|x| \leqslant 1$, and set, respectively,

$$\phi_k = \psi(x/k)\phi, \quad u_k = \psi(x/k)u, \quad k = 1, 2, \ldots.$$

Then $\phi_k \to \phi$ in $C^\infty(\mathbf{R}^n)$, and $u_k \to u$ in $\mathscr{D}'(\mathbf{R}^n)$, as $k \to \infty$.

We remark finally that if $u \in \mathscr{D}'(X)$ and $\psi \in C_c^\infty(X)$ then the distribution $\psi u \in \mathscr{E}'(X)$ is called a *localization of* u.

3.2. Distributions supported at the origin

The extreme case of a distribution with compact support is one whose support is a point. Without loss of generality one can take this to be the origin, and can then assume that $u \in \mathscr{D}'(\mathbf{R}^n)$, since one can set $u = 0$ on $\mathbf{R}^n \backslash \{0\}$. One then has the following result.

Theorem 3.2.1. Suppose that $u \in \mathscr{D}'(\mathbf{R}^n)$ and that $\operatorname{supp} u = \{0\}$. Then there is a nonnegative integer N such that

$$u = \sum_{|\alpha| \leqslant N} c_\alpha \partial^\alpha \delta, \tag{3.2.1}$$

where the c_α are complex numbers.

This will be deduced from

Lemma 3.2.1. If $u \in \mathscr{D}'(\mathbf{R}^n)$ and $\operatorname{supp} u = \{0\}$, then there is an $N \geqslant 0$ such that $\langle u, \phi \rangle = 0$ if $\phi \in C_c^\infty(\mathbf{R}^n)$ and $\partial^\alpha \phi(0) = 0$ for $|\alpha| \leqslant N$.

Proof. Let $\psi \in C_c^\infty(\mathbf{R}^n)$ be such that $\psi = 1$ when $|x| < \frac{1}{2}$ and $\psi = 0$ when $|x| > 1$. Take $\epsilon \in (0, 1)$ and observe that, for any $\phi \in C_c^\infty(\mathbf{R}^n)$, one has

$\phi(x)-\phi(x)\psi(x/\epsilon)=0$ on $|x|<\frac{1}{2}\epsilon$ which is a neighbourhood of supp u. Hence

$$\langle u,\phi\rangle=\langle u(x),\phi(x)\psi(x/\epsilon)\rangle,\quad \phi\in C_c^\infty(\mathbf{R}^n). \tag{3.2.2}$$

Since $x\mapsto\phi(x)\psi(x/\epsilon)$ is supported in the unit ball if $0<\epsilon\leqslant 1$, there is a semi-norm estimate for u which, applied to (3.2.2), gives

$$|\langle u,\phi\rangle|\leqslant C\sum_{|\alpha|\leqslant N}\sup|\partial^\alpha(\phi(x)\psi(x/\epsilon))|,\quad \phi\in C_c^\infty(\mathbf{R}^n). \tag{3.2.3}$$

Assume now that $\partial^\alpha\phi(0)=0$ when $|\alpha|\leqslant N$. If $|\beta|\leqslant N$ then Taylor's theorem, applied to $\partial^\beta\phi$, gives

$$|\partial^\beta\phi(x)|\leq \varepsilon^{N-|\beta|+1}\sum_{|\gamma|=N+1-|\beta|}\sup\{|\partial^\gamma\phi(x)|:|x|<1\}/\gamma!\text{ if }|x|\leqslant\varepsilon.$$

(See the note at the end of this section.) By Leibniz's theorem (cf. (3.5.3))

$$\partial^\alpha(\phi(x)\psi(x/\epsilon))=\sum_{\beta+\gamma=\alpha}\frac{\alpha!}{\beta!\gamma!}\epsilon^{-|\gamma|}\phi^{(\beta)}(x)\psi^{(\gamma)}(x/\epsilon),$$

where $\phi^{(\beta)}(x)=\partial^\beta\phi(x)$, $\psi^{(\gamma)}(x)=\partial^\gamma\psi(x)$. Combining these two results, and taking into account that $\phi(x)\psi(x/\epsilon)=0$ for $|x|>\epsilon$, we deduce that, if $|\alpha|\leqslant N$, then

$$|\partial^\alpha(\phi/x)\psi(x/\epsilon))|\leqslant C_\alpha\epsilon^{N-|\beta|+1}\epsilon^{-|\gamma|}=C_\alpha\epsilon^{N+1-|\alpha|},$$

where C_α is a constant independent of ϵ. Hence one has, from (3.2.3),

$$|\langle u,\phi\rangle|\leqslant C\sum_{|\alpha|\leqslant N}C_\alpha\epsilon^{N+1-|\alpha|}\leqslant B\epsilon$$

for some constant B, and the lemma follows if one makes $\epsilon\to 0$.

Proof of Theorem 3.2.1. With N as in the lemma, and the same ψ, one can write any $\phi\in C_c^\infty(\mathbf{R}^n)$ in the form

$$\phi=\psi(x)\sum_{|\alpha|\leqslant N}\frac{x^\alpha}{\alpha!}\partial^\alpha\phi(0)+\phi'.$$

Then, clearly, $\phi'\in C_c^\infty(\mathbf{R}^n)$ and $\partial^\alpha\phi'(0)$ if $|\alpha|\leqslant N$. So $\langle u,\phi'\rangle=0$, by the lemma, and hence

$$\langle u,\phi\rangle=\sum_{|\alpha|\leqslant N}\frac{1}{\alpha!}\langle u,x^\alpha\psi\rangle\partial^\alpha\phi(0),$$

which is (3.2.1), with $c_\alpha=(-1)^{|\alpha|}\langle u,x^\alpha\psi\rangle/\alpha!$.

Lemma 3.2.1 is a special case of a more general result.

Theorem 3.2.2. Let $X\subset\mathbf{R}^n$ be an open set, and let $u\in\mathscr{E}'(X)$ have order N (necessarily finite). Then $\langle u,\phi\rangle=0$ for all $\phi\in C_c^\infty(X)$ such that $\partial^\alpha\phi(x)=0$ when $x\in$ supp u and $|\alpha|\leqslant N$.

Proof. Write $K = \operatorname{supp} u$, and choose $\delta > 0$ such that

$$K_\delta = \{x \in \mathbf{R}^n : \mathrm{d}(x, K) \leqslant \delta\} \subset X.$$

(K_δ is the closure of the δ-neighbourhood of K, see (1.4.2).) Choose $\rho \in C_c^\infty(\mathbf{R}^n)$ such that $\operatorname{supp} \rho \subset \{|x| \leqslant 1\}$ and $\int \rho \, \mathrm{d}x = 1$, and set

$$\psi_\epsilon(x) = \epsilon^{-n} \int \chi_\epsilon(y) \rho\left(\frac{x-y}{\epsilon}\right) \mathrm{d}y, \tag{3.2.4}$$

where $0 < \frac{1}{4}\epsilon < \delta$, and χ_ϵ is the characteristic function of $K_{2\epsilon}$. Then $\psi_\epsilon = 1$ on K_ϵ, and so

$$|\langle u, \phi \rangle| = 1\langle u, \phi\psi_\epsilon \rangle| \leqslant C \sum_{|\alpha| \leqslant N} \sup |\partial^\alpha(\phi\psi_\epsilon)|,$$

by virtue of a semi-norm estimate for u on K_δ. Arguing exactly as in the proof of lemma 3.2.1, one obtains an estimate $|\langle u, \phi \rangle| \leqslant B\epsilon$ for some constant B, since (3.2.4) implies that $\partial^\gamma \psi_\epsilon = 0(\epsilon^{-|\gamma|})$. The assertion therefore follows when one makes $\epsilon \to 0$.

At first sight, Theorem 3.2.2 seems to imply that one can take $K = \operatorname{supp} u$ in the basic semi-norm estimate (3.2.1), or, equivalently, that if a sequence of functions $(\phi_k)_{1 \leqslant k < \infty}$ converges uniformly to 0 on $\operatorname{supp} u$, and all its derivatives do so as well, then $\langle u, \phi_k \rangle \to 0$ as $k \to \infty$. In general this is false, as is shown by the following example, due to L. Schwartz. Take $X = \mathbf{R}$, and define a distribution u by

$$\langle u, \phi \rangle = \lim_{m \to \infty} \left(\sum_{k=1}^{m} \phi(1/k) - m\phi(0) - \phi'(0) \log m \right), \quad \phi \in C_c^\infty(\mathbf{R}).$$

The support of this is the compact set $\{0\} \cup \{1, \frac{1}{2}, \frac{1}{3}, \ldots\}$. Now one can construct functions $\phi_k \in C_c^\infty(\mathbf{R})$ such that $\phi_k = k^{-1/2}$ for $x \geqslant 1/k$ and $\phi_k = 0$ for $x \leqslant 1/(k+1)$. Then the ϕ_k converge to 0 uniformly as $k \to \infty$, and all their derivatives vanish on the support of u. However

$$\langle u, \phi_k \rangle = k k^{-1/2} = k^{1/2} \to \infty \text{ as } k \to \infty.$$

It must be added, though, that if the boundary of the support of a distribution with compact support is sufficiently regular, for example if it is a smooth surface, then one can take $K = \operatorname{supp} u$ in (3.2.1). This is discussed in [6, pp. 98-100].

Note on Taylor's theorem. If $F(t) \in C^\infty(\mathbf{R})$, then Taylor's formula (2.7.5) gives

$$F(1) = \sum_{k=0}^{N-1} \frac{1}{k!} \partial^k F(0) + \frac{1}{(N-1)!} \int_0^1 (1-t)^{N-1} \partial^N F(t) \, \mathrm{d}t \tag{3.2.5}$$

for any $N \geqslant 1$. Take $f(x) \in C^\infty(\mathbf{R}^n)$, and let $t \in \mathbf{R}$. It is easy to prove by induction that

$$\frac{1}{k!}\,\partial_t^k f(xt) = \frac{1}{k!}\left(\sum_{i=1}^{n} x_i \frac{\partial}{\partial y_i}\right)^k f(y)\bigg|_{y=xt} = \sum_{|\alpha|=k} \frac{x^\alpha}{\alpha!} f^{(\alpha)}(xt),$$

where $f^{(\alpha)}(x)$ stands for $\partial^\alpha f(x)$. Taking F in (3.2.5) to be the function $t \mapsto f(xt)$ one therefore obtains

$$f(x) = \sum_{|\alpha|\leqslant N-1} \frac{x^\alpha}{\alpha!}\,\partial^\alpha f(0) + \sum_{|\alpha|=N} \frac{x^\alpha}{\alpha!} f_\alpha(x), \tag{3.2.6}$$

where

$$f_\alpha(x) = N\int_0^1 (1-t)^{N-1} f^{(\alpha)}(xt)\,\mathrm{d}t. \tag{3.2.7}$$

The remainder term in (3.2.6) can thus be estimated by

$$\sum_{|\alpha|=N} \frac{x^\alpha}{\alpha!} f_\alpha(x) \leqslant |x|^N \sum_{|\alpha|=N} \sup\{|\partial^\alpha f(x')|:|x'|<|x|\}/\alpha! \tag{3.2.8}$$

as one obviously has $|x^\alpha| \leqslant |x|^N$ when $|\alpha| = N$. This is the form of Taylor's theorem used in the proof of Lemma 3.2.1.

Alternatively, one can observe that

$$f_\alpha(x) - \partial^\alpha f(0) = N\int_0^1 (1-t)^{N-1}(f^{(\alpha)}(xt) - f^{(\alpha)}(0))\,\mathrm{d}t = o(1)$$

as $x \to 0$, by the continuity of $f^{(\alpha)} = \partial^\alpha f$. So (3.2.6) and (3.2.7) also imply that

$$f(x) = \sum_{|\alpha|\leqslant N} \frac{x^\alpha}{\alpha!}\,\partial^\alpha f(0) + o(|x|^N), \tag{3.2.9}$$

and this holds for all $f \in C^{N+1}(\mathbf{R}^n)$.

Exercises

3.1. Let $X \subset \mathbf{R}^n$ be an open set, and let F be a closed subset of X. Show that

$\Phi(F) = \{\phi \in C^\infty(x): F \cap \operatorname{supp}\phi \text{ is compact}\}$

is a vector subspace of $C^\infty(X)$. Show further that, if $u \in \mathscr{D}'(X)$, then $\langle u, \phi\rangle$ can be defined unambiguously for all $\phi \in \Phi$ (supp u).

3.2. Let $v \in \mathscr{D}'(\mathbf{R}^+)$ be given by

$$v = \sum_{k=1}^{\infty} \delta_{(1/k)}$$

Find all $u \in \mathscr{D}'(\mathbf{R})$ such that supp $u \subset [0, \infty) = \bar{\mathbf{R}}^+$ and that $u = v$ on $\mathbf{R}^+$.

3.3. Determine all $u \in \mathscr{D}'(\mathbf{R}^2)$ such that

$(x_1 + ix_2)u = 0.$

3.4. Solve the differential equation

$x\partial u - \lambda u = 0,$

where λ is a complex number, in $\mathscr{D}'(\mathbf{R})$.

4 TENSOR PRODUCTS

If X and Y are open sets in $\mathbf{R}^n$ and in $\mathbf{R}^m$, respectively, and if $f \in C^0(X)$ and $g \in C^0(Y)$, say, then a function $f \otimes g \in C^0(X \times Y)$ can be defined by pointwise multiplication,

$$f \otimes g(x,y) = f(x)g(y), \quad (x,y) \in X \times Y.$$

The distribution $f \otimes g \in \mathscr{D}'(X \times Y)$ is then

$$\langle f \otimes g, \chi \rangle = \int f(x)g(y)\chi(x,y)\,\mathrm{d}x\,\mathrm{d}y, \quad \chi \in C_c^\infty(X \times Y).$$

If one takes $\chi = \phi \otimes \psi$ here, one obtains the identity

$$\langle f \otimes g, \phi \otimes \psi \rangle = \langle f, \phi \rangle \langle g, \psi \rangle, \quad \phi \in C_c^\infty(X), \quad \psi \in C_c^\infty(Y). \tag{*}$$

This operation can be extended to distributions; the resulting *tensor product* satisfies (*), and is characterized by this property. Its construction, and basic properties, are the subject of this section. The first step is the derivation of a technical result that is of wider applicability; this will be illustrated by discussing the action of the affine group on $\mathscr{D}'(\mathbf{R}^n)$.

4.1. Test functions which depend on a parameter

We begin with a theorem which provides an essential step in the construction of the tensor product.

Theorem 4.1.1. Let $X \subset \mathbf{R}^n$ and $Y \subset \mathbf{R}^n$ be open sets; let $u \in \mathscr{D}'(X)$, and let $\phi \in C^\infty(X \times Y)$ satisfy the following hypothesis: every point $y' \in Y$ has a neighbourhood $U(y')$ in Y such that the support of $x \mapsto \phi(x,y)$ is contained in a compact set $K = K(y')$ if $y \in U(y')$. Then

$$\langle u(x), \phi(x,y) \rangle \in C^\infty(Y),$$

and

$$\partial^\alpha \langle u(x), \phi(x,y) \rangle = \langle u(x), \partial_y^\alpha \phi(x,y) \rangle \tag{4.1.1}$$

for all multi-indices α.

Note. The notation means that, for each $y \in Y$, u is paired with the test function $\phi(\cdot, y) \in C_c^\infty(X)$.

Proof. Put

$$\psi(y) = \langle u(x), \phi(x,y)\rangle. \tag{4.1.2}$$

We first show that ψ is continuous. Given any point $y' \in Y$, there is a $\delta > 0$ such that $y \in U(y')$ if $|y - y'| < \delta$. Take $h \in \mathbf{R}^m$ with $|h| < \delta$, and set

$$\phi_h(x,y') = \phi(x,y'+h) - \phi(x,y').$$

Then

$$\psi(y'+h) - \psi(y') = \langle u(x), \phi_h(x,y')\rangle.$$

Now supp $\phi_h(\cdot, y') \subset K(y')$, by the hypothesis of the theorem. Again, it follows from the mean value theorem that, for any multi-index α, $\partial_x^\alpha \phi_h(x,y') \to 0$ uniformly as $h \to 0$. Hence $\phi_h(\cdot, y') \to 0$ in $C_c^\infty(X)$ as $h \to 0$, and so $\psi(y'+h) - \psi(y') \to 0$ then, as claimed.

Next, we show that ψ is continuously differentiable. Let e_j be the jth coordinate unit vector in $\mathbf{R}^m$. With y' and δ as above, put

$$\chi_\epsilon(x,y') = \epsilon^{-1}(\phi(x,y'+\epsilon e_j) - \phi(x,y')) - (\partial/\partial y_j)\phi(x,y').$$

where $0 < \epsilon < \delta$. Then, by (4.1.2)

$$\begin{aligned} &\epsilon^{-1}(\psi(y'+\epsilon e_j) - \psi(y')) - \langle u(x), (\partial(\partial y_j)\phi(x,y')\rangle \\ &\qquad = \langle u(x), \chi_\epsilon(x,y')\rangle. \end{aligned}$$

The choice of δ implies that supp $\chi_\epsilon(\cdot, y') \subset K(y')$. Taylor's theorem shows that, for each α, the $\partial_x^\alpha \chi(x,y')$ converge uniformly to zero as $\epsilon \to 0$. Hence the partial derivative $\partial_j \psi$ exists and (writing y instead of y')

$$\partial_j \psi(y) = \langle u(x), (\partial(\partial y_j)\phi(x,y)\rangle. \tag{4.1.3}$$

As $(\partial/\partial y_j)\phi$ satisfies the hypothesis of the theorem if ϕ does so, it follows by what has already been proved that $\partial_j \psi$ is continuous. This being the case for $j = 1, \ldots, n$, one can conclude that $\psi \in C^1(Y)$.

It now follows by induction on the order that $\psi \in C^\infty(Y)$, and, likewise, (4.1.1) follows by induction from (4.1.3). So the theorem is proved.

Two special cases are worth noting separately.

Corollary 4.1.1. Let $u \in \mathscr{D}'(X)$ and $\phi \in C_c^\infty(X \times Y)$. Then $\langle u(x), \phi(x,y)\rangle \in C_c^\infty(Y)$.

Proof. One can take $U(y) = Y$ for all $y \in Y$, and $K(y)$ equal to the projection of supp ϕ on X. Also, supp ψ (ψ as in (4.1.2)) is a subset of the projection of supp ϕ on Y, which is a compact set.

Corollary 4.1.2. Let $u \in \mathscr{E}'(X)$ and $\phi \in C^\infty(X \times Y)$. Then $\psi = \langle u(x), \phi(x,y)\rangle \in C^\infty(Y)$.

Proof. By definition, the function in question is

$$\psi(y) = \langle u(x), \rho(x)\phi(x,y)\rangle,$$

where the cut-off function $\rho \in C_c^\infty(X)$ is equal to unity on a neighbourhood of $\operatorname{supp} u$. So one can again take $U = Y$ for all $y \in Y$, and $K = \operatorname{supp} \rho$.

4.2. Affine transformations

We digress to give two applications of Theorem 4.1.1. An affine transformation of $\mathbf{R}^n$ is a map $x \mapsto Ax - h$, where $h \in \mathbf{R}^n$ and A is a non-singular $n \times n$ matrix. If f is a function defined on $\mathbf{R}^n$ then one can form the composite function $x \mapsto f(Ax - h)$. A similar operation can be performed on distributions. We begin with translation.

Let $h \in \mathbf{R}^n$ be a fixed vector. The translate $\tau_h f$ of a function f is, by definition, the function

$$\tau_h f(x) = f(x - h), \quad x \in \mathbf{R}^n. \tag{4.2.1}$$

For the distribution determined by $\tau_h f$ one has

$$\langle \tau_h f, \phi \rangle = \int f(x-h)\phi(x)\,dx = \int f(x)\phi(x+h)\,dx = \langle f, \tau_{-h}\phi \rangle \tag{4.2.2}$$

where $\phi \in C_c^\infty(\mathbf{R}^n)$; one may here take f to be continuous, or locally integrable. It is a trivial exercise to verify that the map $\phi \mapsto \tau_{-h}\phi$ is a continuous map $C_c^\infty(\mathbf{R}^n) \to C_c^\infty(\mathbf{R}^n)$ in the sense of Definition 2.8.1. So one can use the duality principle to define the translate of a distribution:

Definition 4.2.1. Let $u \in \mathscr{D}'(\mathbf{R}^n)$, and $h \in \mathbf{R}^n$. The translate $\tau_h u$ of u is the distribution

$$\begin{aligned} \langle \tau_h u, \phi \rangle &= \langle u, \tau_{-h}\phi \rangle \\ &= \langle u(x), \phi(x+h) \rangle, \quad \phi \in C_c^\infty(\mathbf{R}^n). \end{aligned} \tag{4.2.3}$$

Note. We shall sometimes write $u(x-h)$ for $\tau_h u$.

Theorem 4.2.1. (i) When $u \in C^0(\mathbf{R}^n)$, then $\tau_h u$ is equal to the function (4.2.1).

(ii) The map $\tau_h : \mathscr{D}'(\mathbf{R}^n) \to \mathscr{D}'(\mathbf{R}^n)$ is sequentially continuous.

(iii) Translation commutes with differentiation.

(iv) $h \mapsto \tau_h u$ is a C^∞ function $\mathbf{R}^n \to \mathscr{D}'(\mathbf{R}^n)$, and

$$(\partial/\partial h_i)\tau_h u = -\partial_i \tau_h u, \quad i = 1, \ldots, n. \tag{4.2.4}$$

Proof. (i) This follows from (4.2.2) and (4.2.1).

(ii) This is also obvious. (See also Theorem 2.8.1.)

(iii) This is the dual of the identity

$$\tau_{-h}\partial_i \phi = \partial_i \phi(x+h) = \partial_i \tau_{-h}\phi.$$

(iv) The statement means that $h \mapsto \langle \tau_h u, \phi \rangle$ is C^∞ for all $\phi \in C_c^\infty(\mathbf{R}^n)$. It is at this point that one can appeal to Theorem 4.1.1. Indeed, the support of

$x \mapsto \phi(x+h)$ is in the δ-neighbourhood of

$$\operatorname{supp} \tau_{-h}\phi = \{x : x+h \in \operatorname{supp} \phi\}$$

when $|h| < \delta$, so that the hypothesis of this theorem holds. Also, (4.2.4) is a consequence of (4.1.1) and (4.2.3).

Again, let $A = (a_{ij})$ be an invertible $n \times n$ matrix, with real entries; then one can define a map $x \mapsto Ax$ by

$$(Ax)_i = \sum_{j=1}^{n} a_{ij} x_j, \quad i = 1, \ldots, n,$$

which is of course a homomorphism $\mathbf{R}^n \to \mathbf{R}^n$. If (say) $f \in C^0(\mathbf{R}^n)$, we write

$$A^*f(x) = f(Ax), \quad x \in \mathbf{R}^n, \tag{4.2.5}$$

for the composite function. For the corresponding distribution one has

$$\begin{aligned} \langle A^*f, \phi \rangle &= \int f(Ax)\phi(x)\, dx \\ &= |\det A|^{-1} \int f(x)\phi(A^{-1}x)\, dx = |\det A|^{-1} \langle f, (A^{-1})^*\phi \rangle \end{aligned} \tag{4.2.6}$$

where $\phi \in C_c^\infty(\mathbf{R}^n)$, by a change of the variable of integration. It is again straightforward to verify that Theorem 2.8.1 applies, and so one makes the following definition:

Definition 4.2.2. Let $u \in \mathscr{D}'(\mathbf{R}^n)$, and let A be a real invertible $n \times n$ matrix. Then the distribution A^*u is defined to be

$$\langle A^*u, \phi \rangle = |\det A|^{-1} \langle u, (A^{-1})^*\phi \rangle, \quad \phi \in C_c^\infty(\mathbf{R}^n). \tag{4.2.7}$$

Before stating the properties of A^*u as a theorem, we recall that a matrix $A = (a_{ij})_{1 < i,j < n}$ can be represented by a point in $\mathbf{R}^{n^2}$, whose coordinates are the a_{ij}, arranged in some definite order. The set of invertible matrices is characterized by the condition $\det A \neq 0$, and so is an open set; in effect, it is the general linear group $GL(n, \mathbf{R})$. So one can consider $A \mapsto A^*u$ as a function $GL(n, \mathbf{R}) \to \mathscr{D}'(\mathbf{R}^n)$.

Theorem 4.2.2. Let u and A be as in Definition 4.2.2. Then (i) A^*u is equal to the function (4.2.5) if $u \in C^0(\mathbf{R}^n)$; (ii) the map $u \mapsto A^*u$ is a sequentially continuous map $\mathscr{D}'(\mathbf{R}^n) \to \mathscr{D}'(\mathbf{R}^n)$; (iii) for fixed u, the map $A \mapsto A^*u$ is a C^∞ function $GL(n, \mathbf{R}) \to \mathscr{D}'(\mathbf{R}^n)$.

Proof. Both (i) and (ii) are obvious, from (4.2.7) and (4.2.6). As to (iii), it amounts to showing that $A \mapsto \langle A^*u, \phi \rangle$ is a C^∞ function on $GL(n, \mathbf{R})$ for each

$\phi \in C_c^\infty(\mathbf{R}^n)$. This follows from the definition, (4.2.7). For, if A is restricted to a relatively compact open subset of $GL(n, \mathbf{R})$, then it is evident that

$$\text{supp}\,(A^{-1})^*\phi = \{x: A^{-1}x \in \text{supp}\,\phi\} = A\,(\text{supp}\,\phi)$$

is contained in a compact set. So the hypothesis of Theorem 4.1.1 is satisfied, and we are done. The theorem is proved.

The derivatives of A^*u with respect to the entries of A can be computed by the chain rule; for the details, see Exercise 4.2.

We note two special cases, which occur frequently. If $A = -I$, where I is the identity, one obtains the *reflection* map which is usually written as $\check{u}$; so for a function one has

$$\check{\phi}(x) = \phi(-x), \quad x \in \mathbf{R}^n, \tag{4.2.8}$$

and for a distribution,

$$\langle \check{u}, \phi \rangle = \langle u, \check{\phi} \rangle, \quad \phi \in C_c^\infty(\mathbf{R}^n). \tag{4.2.9}$$

Thus one can speak of even and odd distributions; for these one has $\check{u} = u$ and $\check{u} = -u$, respectively. For example, $\partial^\alpha \delta$ is even or odd according as $|\alpha|$ is even or odd, respectively.

If $A = tI$, where t is a positive real number, one has a *dilation*. Let us write u_t for A^*u in this case. Then (4.2.7) gives

$$\langle u_t, \phi \rangle = t^{-n} \langle u(x), \phi(x/t) \rangle, \quad \phi \in C_c^\infty(\mathbf{R}^n). \tag{4.2.10}$$

A function or distribution u is said to be (positively) *homogeneous* of degree λ, where λ is a complex number, if

$$u_t = t^\lambda u, \quad t > 0. \tag{4.2.11}$$

For a function, this is the usual convention, $f(xt) = t^\lambda f(x)$ for all $x \in \mathbf{R}^n$ and $t > 0$; for a distribution it means, by (4.2.10), with t replaced by $1/t$, that

$$\langle u, \phi \rangle = t^{n+\lambda} \langle u(x), \phi(xt) \rangle, \quad t > 0, \quad \phi \in C_c^\infty(\mathbf{R}^n). \tag{4.2.12}$$

The Dirac distribution on $\mathbf{R}^n$ is positively homogeneous of degree $-n$.

4.3. The tensor product of distributions

We now return to the main subject of this section. We need one more preliminary result.

Theorem 4.3.1. If $X \subset \mathbf{R}^n$ and $Y \subset \mathbf{R}^m$ are open sets, then the subspace of $C_c^\infty(X \times Y)$ generated by functions of the form $\phi \otimes \psi$, where $\phi \in C_c^\infty(X)$ and $\psi \in C_c^\infty(Y)$, is dense in $C_c^\infty(X \times Y)$.

Proof. It has to be shown that, if $\chi \in C_c^\infty(X \times Y)$, then there is a sequence of functions

$$\chi_m = \sum_{j=1}^{m} \phi_j \otimes \psi_j, \quad \phi_j \in C_c^\infty(X), \quad \psi_j \in C_c^\infty(Y), \quad m = 1, 2, \ldots \tag{4.3.1}$$

which converges to χ in $C_c^\infty(X \times Y)$. By a partition of unity, the proof of this can be reduced to the case where the support of χ is contained in a cube. This, in turn, follows from the case where the support of χ is a subset of the unit cube in $\mathbf{R}^n \times \mathbf{R}^m$. But then it is clear that the theorem is implied by the following lemma.

Lemma 4.3.1. Let $N > 1$ be an integer, let $I = (0, 1)^N$ be the unit cube in $\mathbf{R}^N$, and suppose that $\phi \in C_c^\infty(I)$. Then one can find functions $\psi_{jk} \in C_c^\infty(0, 1)$ where $j = 1, 2, \ldots$ and $k = 1, \ldots, N$, such that the sequence

$$\phi_m(z) = \sum_{j=1}^{m} \psi_{j1}(z_1) \ldots \psi_{jN}(z_N), \quad m = 1, 2, \ldots \tag{4.3.2}$$

converges to ϕ in $C_c^\infty(I)$.

Proof. Extend ϕ to $\bar{I}$ by setting $\phi = 0$ on ∂I, and define a periodic function $\tilde{\phi}$ on $\mathbf{R}^N$ by setting $\tilde{\phi}(z) = \phi(z')$ when $z \equiv z' (\text{mod } \mathbf{Z}^N)$. (Here, $\mathbf{Z}$ stands for the rational integers, and $\mathbf{Z}^N$ is the lattice in $\mathbf{R}^N$ consisting of points with integer coordinates.) One can expand $\tilde{\phi}$ as a Fourier series,

$$\tilde{\phi} = \sum_{g \in \mathbf{Z}^N} \hat{\phi}_g \exp(2\pi i g \cdot z)$$

where

$$\hat{\phi}_g = \int_I \phi(z) \exp(-2\pi i g \cdot z)\, dz, \quad g \in \mathbf{Z}^N.$$

It is well known that this series converges to $\tilde{\phi}$ in $C^\infty(\mathbf{R}^N)$, and so to ϕ in $C^\infty(\bar{I})$. We assume this without proof, and remark only that the proof follows from the fact, easily proved by partial integration, that $|g|^M \hat{\phi}_g \to 0$ as $|g| \to \infty$ for any $M \geqslant 0$.

Now as ϕ is supported in I, there is a $\delta > 0$ such that supp $\phi \subset [\delta, 1 - \delta]^N$. Choose $\rho \in C_c^\infty(0, 1)$ such that $\rho = 1$ on $(\frac{1}{2}\delta, 1 - \frac{1}{2}\delta)$, and set

$$\phi_m = \sum_{|g_1| \leqslant m, \ldots |g_N| \leqslant m} \hat{\phi}_g \prod_{k=1}^{N} \rho(z_k) \exp(2\pi i g_k z_k), \quad m = 0, 1, \ldots$$

It is clear from Leibniz's theorem and the convergence of the Fourier series to ϕ in $C^\infty(\bar{I})$ that these functions, which are of the form (4.3.2), converge to ϕ in $C_c^\infty(I)$, and so the lemma is proved.

We are now ready for the final step.

Theorem 4.3.2. Let $X \subset \mathbf{R}^n$ and $Y \subset \mathbf{R}^m$ be open sets, and suppose that $u \in \mathscr{D}'(X)$ and $v \in \mathscr{D}'(Y)$. Then there is a unique element of $\mathscr{D}'(X \times Y)$, called the tensor product of u and v, and written as $u \otimes v$, such that

$$\langle u \otimes v, \phi \otimes \psi \rangle = \langle u, \phi \rangle \langle v, \psi \rangle, \quad \phi \in C_c^\infty(X), \quad \psi \in C_c^\infty(Y). \tag{4.3.3}$$

Proof. The identity (4.3.3) determines a linear form on the subspace of $C_c^\infty(X \times Y)$ generated by elements of the form $\phi \otimes \psi$. Indeed, if

$$\chi = \sum_{j=1}^{m} \phi_j \otimes \psi_j, \quad \phi_j \in C_c^\infty(X), \quad \psi_j \in C_c^\infty(Y), \tag{4.3.4}$$

then it follows from linearity and (4.3.3) that

$$\langle u \otimes v, \chi \rangle = \sum_{j=1}^{m} \langle u, \phi_j \rangle \langle v, \psi_j \rangle. \tag{4.3.5}$$

Theorem 4.3.1 therefore implies uniqueness, and we only have to prove the existence of the tensor product. Now the second member of (4.3.5) is

$$\left\langle v, \sum_{j=1}^{m} \langle u, \phi_j \rangle \psi_j \right\rangle = \langle v(y), \langle u(x), \chi(x,y) \rangle \rangle.$$

If $\phi \in C_c^\infty(X \times Y)$, then $\langle u(x), \phi(x,y) \rangle \in C_c^\infty(Y)$, by Corollary 4.1.1. So one can define a linear form $u \otimes v$ on $C_c^\infty(X \times Y)$ by setting

$$\langle u \otimes v, \phi \rangle = \langle v(y), \langle u(x), \phi(x,y) \rangle \rangle, \quad \phi \in C_c^\infty(X \times Y). \tag{4.3.6}$$

This reduces to (4.3.5) when ϕ is of the form (4.3.4), and in particular it satisfies (4.3.3). So it only remains to prove that (4.3.6) is a distribution. Let $K \subset X \times Y$ be a compact set, and let

$$K_1 = \{x \in X\text{: there is a } y \text{ such that } (x,y) \in K\}$$
$$K_2 = \{y \in Y\text{: there is an } x \text{ such that } (x,y) \in K\}$$

be its projections on X and on Y, respectively. Take $\phi \in C_c^\infty(X \times Y)$ such that $\operatorname{supp} \phi \subset K$, and put $\langle u(x), \phi(x,y) \rangle = g(y)$. Clearly, $\operatorname{supp} g \subset K_2$; so there is a semi-norm estimate for v which gives

$$|\langle v, g \rangle| \leqslant C \sum_{|\beta| \leqslant M} \sup |\partial^\beta g|,$$

for some constants $C, M \geqslant 0$. By (4.1.1),

$$\partial^\beta g(y) = \langle u(x), \partial_y^\beta \phi(x,y) \rangle.$$

But the support of $x \mapsto \partial_y^\beta \phi(x,y)$ is a subset of K_1, so there is a semi-norm estimate for u which gives

$$|\partial^\beta g(y)| \leqslant C' \sum_{|\alpha| \leqslant N} \sup \{|\partial_x^\alpha \partial_y^\beta \phi(x,y)| : x \in K_1\}$$

$$\leqslant C' \sum_{|\alpha| \leqslant N} \sup |\partial_x^\alpha \partial_y^\beta \phi|.$$

Combining these two inequalities, one gets a semi-norm estimate for the linear form (4.3.6), and so we are done.

The next theorem lists the properties of the tensor product.

Theorem 4.3.3. (i) The tensor product $u \otimes v$ can be computed as

$$\begin{aligned}\langle u \otimes v, \phi\rangle &= \langle v(y), \langle u(x), \phi(x,y)\rangle\rangle \\ &= \langle u(x), \langle v(y), \phi(x,y)\rangle\rangle, \quad \phi \in C_c^\infty(X \times Y).\end{aligned} \tag{4.3.7}$$

(ii) The support of $u \otimes v$ is the product of supp u and supp v.

(iii) If α is an n-multi-index and β is an m-multi-index, then

$$\partial_x^\alpha \partial_y^\beta u \otimes v = \partial_x^\alpha u \otimes \partial_y^\beta v. \tag{4.3.8}$$

(iv) The tensor product is a separately continuous bilinear form on $\mathscr{D}'(X) \times \mathscr{D}'(Y)$.

Proof. (i) We have defined $u \otimes v$ by means of (4.3.7). The proof of Theorem 4.3.2, with u and v interchanged, shows that $\phi \mapsto \langle u(y), \langle v(x), \phi(x,y)\rangle\rangle$ is a distribution; it is also evident that this distribution satisfies (4.3.3). So the identity (4.3.7) follows from the uniqueness of the tensor product.

(ii) It is clear from (4.3.7) that $\langle u \otimes v, \phi\rangle = 0$ if supp $\phi \subset (X \setminus \text{supp}\, u) \times Y$, or if supp $\phi \subset X \times (Y \setminus \text{supp}\, v)$, so also if the support of ϕ is disjoint from supp $u \times$ supp v; hence

$$\text{supp}\, u \otimes v \subset \text{supp}\, u \times \text{supp}\, v.$$

On the other hand, if $x \in \text{supp}\, u$ and $y \in \text{supp}\, y$, then one can find $\phi \in C_c^\infty(X)$ and $\psi \in C_c^\infty(Y)$ such that $\langle u, \phi\rangle \neq 0$ and $\langle v, \psi\rangle \neq 0$, so it follows from (4.3.3) that $(x,y) \in \text{supp}\, u \otimes v$, which gives the opposite inclusion; so one has

$$\text{supp}\, u \otimes v = \text{supp}\, u \times \text{supp}\, v. \tag{4.3.9}$$

Both (iii) and (iv) are immediate consequences of (4.3.7).

Remark. The tensor product of any finite set of distributions is constructed in the same way. It has similar properties and is, in addition, associative.

As an example, we observe that $\mathbf{R}^n \cong \mathbf{R} \times \ldots \times \mathbf{R}$, and compute $\delta(x_1) \otimes \ldots \otimes \delta(x_n)$. The obvious extension of (4.3.7) to this case gives at once

$$\delta(x) = \delta(x_1) \otimes \ldots \otimes \delta(x_n). \tag{4.3.10}$$

Since $\partial H = \delta$ in $\mathscr{D}'(\mathbf{R})$ one can thus deduce from (4.3.8) that

$$\partial_1 \ldots \partial_n (H(x_1) \otimes \ldots \otimes H(x_n)) = \delta. \tag{4.3.11}$$

Another application of Theorem 4.3.2 is the following:

Theorem 4.3.4. If $u \in \mathscr{D}'(\mathbf{R}^n)$ then $\partial_n u = 0$ if and only if

$$u = v(x') \otimes 1(x_n), \tag{4.3.12}$$

where $x' = (x_1, \ldots, x_{n-1}) \in \mathbf{R}^{n-1}$, $v \in \mathscr{D}'(\mathbf{R}^{n-1})$, and $1(t)$ is the constant function, equal to unity, on $\mathbf{R}$.

Proof. It is immediate from (4.3.8) that (4.3.12) implies that $\partial_n u = 0$. To prove the converse, choose some $\chi \in C_c^\infty(\mathbf{R})$ such that

$$\int \chi(t)\, \mathrm{d}t = 1, \tag{4.3.13}$$

and define v by

$$\langle v, \psi \rangle = \langle u, \psi(x') \otimes \chi(x_n) \rangle, \quad \psi \in C_c^\infty(\mathbf{R}^{n-1}). \tag{4.3.14}$$

Obviously, v is a member of $\mathscr{D}'(\mathbf{R}^{n-1})$. One now has

$$\langle v \otimes 1(x_n), \phi \rangle = \left\langle v(x'), \int \phi(x', t)\, \mathrm{d}t \right\rangle$$

$$= \left\langle u(x', x_n), \chi(x_n) \int \phi(x', t)\, \mathrm{d}t, \right\rangle \quad \phi \in C_c^\infty(\mathbf{R}^n).$$

So

$$\langle u, \phi \rangle - \langle v \otimes 1(x_n), \phi \rangle = \langle u, \tilde{\phi} \rangle$$

where

$$\tilde{\phi}(x) = \phi(x) - \chi(x_n) \int \phi(x', t)\, \mathrm{d}t.$$

But

$$\int \tilde{\phi}(x', x_n)\, \mathrm{d}x_n = 0, \quad x' \in \mathbf{R}^{n-1}$$

by (4.3.13), so that

$$\int_{x_n}^\infty \tilde{\phi}(x', t)\, \mathrm{d}t \in C_c^\infty(\mathbf{R}^n).$$

Hence

$$\langle u, \phi \rangle - \langle v \otimes 1(x_n), \phi \rangle = \left\langle \partial_n u, \int_{x_n}^\infty \tilde{\phi}(x', t)\, \mathrm{d}t \right\rangle = 0$$

since $\partial_n u = 0$ by hypothesis, and we are done.

Exercises

4.1. Let $u \in \mathscr{D}'(\mathbf{R}^n)$ and $h \in \mathbf{R}^n$. Show that

$$\lim_{\epsilon \to 0} \frac{1}{\epsilon} (\tau_{-\epsilon h} u - u) = \sum_{i=1}^n h_i \partial_i u$$

in $\mathscr{D}'(\mathbf{R}^n)$. (So the derivatives of u are the limits of the usual difference quotients, but in the distribution topology.)

4.2. Let $A = (a_{ij})$ be an invertible matrix, and let $u \in \mathscr{D}'(\mathbf{R}^n)$. Show that the derivatives of u can be computed by the chain rule,

$$\partial_i(A^*u) = A^*\left(\sum_{j=1}^{n} a_{ji}\partial_j u\right), \qquad i = 1, \ldots, n.$$

4.3. Suppose that $u \in \mathscr{D}'(\mathbf{R}^n)$ is homogeneous of degree λ, where λ is a complex number. (i) Show that $\partial^\alpha u$ is homogeneous of degree $\lambda - |\alpha|$. (ii) Show that Euler's equation holds,

$$\sum_{i=1}^{n} x_i \partial_i u = \lambda u.$$

(iii) List all distributions on $\mathbf{R}$ which are homogeneous of degree λ.

4.4. Let $X \subset \mathbf{R}^n$ and $Y \subset \mathbf{R}^m$ be open sets, and put $Z = X \times Y$. Let $u \in \mathscr{D}'(X)$ be given, and define a linear form on $C_c^\infty(Z)$ by

$$\langle v, \phi \rangle = \int \langle u(x), \phi(x, y) \rangle\rangle \, dy, \qquad \phi \in C_c^\infty(Z).$$

Show that $v \in \mathscr{D}'(Z)$. Also, prove that

$$\left\langle u(x), \int \phi(x, y)\, dy \right\rangle = \int \langle u(x), \phi(x, y) \rangle \, dy = \langle v, \phi \rangle$$

for all $\phi \in C_c^\infty(Z)$.

4.5. Show that if $u \in \mathscr{D}'(\mathbf{R}^n)$, then $x_n u = 0$ if and only if

$u = v(x') \otimes \delta(x_n)$

where $v \in \mathscr{D}'(\mathbf{R}^{n-1})$. (The notation is as in Theorem 4.3.4.)

4.6. Let $k \leqslant n-1$ be a positive integer, and write $x \in \mathbf{R}^n$ as (x', x''), where $x' = (x_1, \ldots, x_k)$ and $x'' = (x_{k+1}, \ldots, x_n)$. Let $u \in \mathscr{E}'(\mathbf{R}^n)$, and suppose that $\operatorname{supp} u \subset \{x : x'' = 0\}$. Show that

$$u = \sum_{\beta \in B} u_\beta(x') \otimes (\partial'')^\beta \delta(x'') \qquad (*)$$

where B is a finite set of multi-indices, $\delta(x'')$ is the Dirac distribution on $\mathbf{R}^{n-k}$, and the u_β are members of $\mathscr{E}'(\mathbf{R}^k)$. How does (*) have to be modified when $u \in \mathscr{D}'(\mathbf{R}^n)$?

(Suggestion: Consider the bilinear form $(\phi, \psi) \mapsto \langle u, \phi \otimes \psi \rangle$ where $\phi \in C_c^\infty(\mathbf{R}^k)$ and $\psi \in C_c^\infty(\mathbf{R}^{n-k})$; what does this become when ϕ, respectively ψ, is fixed?)

5 CONVOLUTION

The convolution of two functions f and g on $\mathbf{R}^n$ is the function

$$f * g(x) = \int f(y)g(x-y)\,\mathrm{d}y = \int f(x-y)g(y)\,\mathrm{d}y, \quad x \in \mathbf{R}^n. \tag{$*$}$$

Some hypotheses on f and g are needed to ensure that the integral exists. For example, one can assume that both functions are continuous, and that one of them has compact support. Then $f*g$ exists and is continuous, and so determines a distribution which can readily be seen to be

$$\langle f * g, \phi \rangle = \int f(x)g(y)\phi(x+y)\,\mathrm{d}x\,\mathrm{d}y, \quad \phi \in C_c^\infty(\mathbf{R}^n). \tag{$**$}$$

Taking this as a model, one can define the convolution of two distributions.

Convolution has many interesting applications. One of these is a proof of the important fact that C_c^∞ is dense in $\mathscr{D}'$; another one is an easy proof of the structure theorem, which says that the restriction of a distribution to a bounded open set is a derivative of some finite order of a continuous function.

The vector space structure of $\mathbf{R}^n$ plays an important part in this section. The following notation will be used. For any set $A \subset \mathbf{R}^n$, one puts

$$-A = \{x : -x \in A\}.$$

If B is another set in $\mathbf{R}^n$ then the *vector sum* and *vector difference* of A and B are, respectively,

$$A + B = \{x : x = y + z, y \in A, z \in B\},$$
$$A - B = A + (-B) = \{x : x = y - z, y \in A, z \in B\}.$$

5.1. The convolution of two distributions

Formally, the equation ($**$) can be written as

$$\langle u * v, \phi \rangle = \langle u(x) \otimes v(y), \phi(x+y) \rangle, \quad \phi \in C_c^\infty(\mathbf{R}^n).$$

When both u and v have compact support, then so has $u \otimes v$, by Theorem 4.3.3 (ii), and the second member is well defined. But in general, it may not exist, because $(x,y) \mapsto \phi(x+y)$ does not have compact support. A simple way to overcome this difficulty is to assume that one of the distributions u and v has compact support.

Let us then assume that $u\in\mathscr{E}'(\mathbf{R}^n)$ and $v\in\mathscr{D}'(\mathbf{R}^n)$. Choose a cut-off function $\rho\in C_c^\infty(\mathbf{R}^n)$ such that $\rho=1$ on a neighbourhood of the support of u. Now

$$\operatorname{supp}(\rho(x)\phi(x+y)) = \operatorname{supp}\rho\times(\operatorname{supp}\phi-\operatorname{supp}\rho) \tag{5.1.1}$$

is a compact subset of $\mathbf{R}^n\times\mathbf{R}^n$ if $\phi\in C_c^\infty(\mathbf{R}^n)$. One can therefore define a linear form $u*v$ on $C_c^\infty(\mathbf{R}^n)$ by setting

$$\langle u*v,\phi\rangle = \langle u(x)\otimes v(y), \rho(x)\phi(x+y)\rangle,\quad \phi\in C_c^\infty(\mathbf{R}^n). \tag{5.1.2}$$

This is independent of the choice of cut-off function. For, if $\sigma\in C_c^\infty(\mathbf{R}^n)$ and $\sigma=1$ on a neighbourhood of supp u, then

$$\langle u(x)\otimes v(y), \rho(x)\phi(x+y)-\sigma(x)\phi(x+y)\rangle = 0$$

since $(\rho(x)-\sigma(x))\phi(x+y)=0$ on a neighbourhood of

$$\operatorname{supp}u\times\operatorname{supp}v = \operatorname{supp}u\otimes v.$$

If we now fix the cut-off function, and take $\phi\in C_c^\infty(K)$, where $K\subset\mathbf{R}^n$ is a compact set, then it follows from (5.1.1) that the support of $\rho(x)\phi(x+y)$ is contained in $\operatorname{supp}\rho\times(K-\operatorname{supp}\rho)$, which is a fixed compact subset of $\mathbf{R}^n\times\mathbf{R}^n$. So one has a semi-norm estimate for $u\otimes v$, and this obviously implies that (5.1.2) yields a distribution. By definition, this is the convolution of u and v:

Theorem 5.1.1. Let $u\in\mathscr{E}'(\mathbf{R}^n)$ and $v\in\mathscr{D}'(\mathbf{R}^n)$. Then (5.1.2) defines a distribution, which is called the convolution of u and v. It can be computed as follows:

$$\begin{aligned}\langle u*v,\phi\rangle &= \langle v(y), \langle u(x), \phi(x+y)\rangle\rangle\\ &= \langle u(x), \langle v(y), \phi(x+y)\rangle\rangle,\quad \phi\in C_c^\infty(\mathbf{R}^n).\end{aligned} \tag{5.1.3}$$

Proof. We have already shown that $u*v$ is a distribution. Also, Theorem 4.3.3 (i) and (5.1.2) give

$$\begin{aligned}\langle u*v,\phi\rangle &= \langle v(y), \langle u(x), \rho(x)\phi(x+y)\rangle\rangle\\ &= \langle u(x), \rho(x)\langle v(y), \phi(x+y)\rangle\rangle,\end{aligned}$$

and this is (5.1.3), by virtue of the Remark following Theorem 3.1.2.

Remark. One can obviously interchange the parts played by u and v in this construction; so the convolution is commutative.

We shall now derive the properties of the convolution of two distributions. We begin with an outer bound for the support.

Theorem 5.1.2. Let u and v be two distributions on $\mathbf{R}^n$, at least one of which has compact support. Then

$$\operatorname{supp}u*v\subset\operatorname{supp}u+\operatorname{supp}v. \tag{5.1.4}$$

Proof. We first show that if $A \subset \mathbf{R}^n$ is compact, and $B \subset \mathbf{R}^n$ is closed, then the vector sum $A+B$ is a closed set. To see this, consider a sequence of points $(x_j)_{1 \leq j < \infty} \in A+B$ which converges to $x \in \mathbf{R}^n$; we have to show that $x \in A+B$. Now $x_j = y_j + z_j$, say, where $y_j \in A$ and $z_j \in B$. As A is compact, the sequence (y_j) has a convergent subsequence, whose limit is a point $y \in A$. Going over to this subsequence, one obtains $z_j = x_j - y_j \to x - y = z$, say, as $j \to \infty$. But $z \in B$ since B is closed, and so $x = y + z$ with $y \in A$ and $z \in B$, whence $x \in A+B$, which proves the claim.

From this result, and the hypothesis of the theorem, it follows that $\operatorname{supp} u + \operatorname{supp} v$ is closed. So (5.1.4) will be proved once it is shown that the restriction of $u * v$ to the open set $\mathbf{R}^n \setminus (\operatorname{supp} u + \operatorname{supp} v)$ is 0. But this is immediate from the definition (5.1.2); if the support of $\phi \in C_c^\infty(\mathbf{R}^n)$ is disjoint from $\operatorname{supp} u + \operatorname{supp} v$, then the support of $(x,y) \mapsto \phi(x+y)$ is disjoint from $\operatorname{supp} u \otimes v = \operatorname{supp} u \times \operatorname{supp} v$, since this equation shows that $(x,y) \in \operatorname{supp} u \otimes v$ implies that $x+y \in \operatorname{supp} u + \operatorname{supp} v$. So the theorem is proved.

We next list some simple rules for manipulating convolutions:

$$\partial_j(u * v) = \partial_j u * v = u * \partial_j v, \quad j = 1, \ldots, n; \tag{5.1.5}$$

$$\tau_h(u * v) = \tau_h u * v = u * \tau_h v, \quad h \in \mathbf{R}^n. \tag{5.1.6}$$

As these are all easily derived from (5.1.3), the proofs are left to the reader. Another elementary but important identity is

$$\delta * u = u, \quad u \in \mathscr{D}'(\mathbf{R}^n). \tag{5.1.7}$$

This also follows from (5.1.3):

$$\langle \delta * u, \phi \rangle = \langle u(x), \langle \delta(y), \phi(x+y) \rangle\rangle = \langle u(x), \phi(x) \rangle.$$

It is obvious that $u * v$ is a bilinear form on $C_c^\infty(\mathbf{R}^n)$. The next theorem asserts that it is separately sequentially continuous with an appropriate definition of convergence in $\mathscr{E}'(\mathbf{R}^n)$.

Theorem 5.1.3. (i) Suppose that $u \in \mathscr{E}'(\mathbf{R}^n)$ and that the sequence $(v_j)_{1 \leq j < \infty}$ converges to $v \in \mathscr{D}'(\mathbf{R}^n)$ in $\mathscr{D}'(\mathbf{R}^n)$. Then $u * v_j \to u * v$ in $\mathscr{D}'(\mathbf{R}^n)$, as $j \to \infty$.

(ii) Suppose that $u \in \mathscr{D}'(\mathbf{R}^n)$, that the sequence $(v_j)_{1 \leq j < \infty}$ converges to $v \in \mathscr{D}'(\mathbf{R}^n)$ in $\mathscr{D}'(\mathbf{R}^n)$, and that the supports of the v_j are contained in a fixed compact set K. Then $u * v_j \to u * v$ in $\mathscr{D}'(\mathbf{R}^n)$, as $j \to \infty$.

Proof. (i) If $u \in \mathscr{E}'(\mathbf{R}^n)$ and $\phi \in C_c^\infty(\mathbf{R}^n)$, then $\langle u(x), \phi(x+y) \rangle$ has compact support, and is C^∞ by Corollary 4.1.2. Hence it follows from (5.1.3) that, for all $\phi \in C_c^\infty(\mathbf{R}^n)$,

$$\begin{aligned} \lim_{j \to \infty} \langle u * v_j, \phi \rangle &= \lim_{j \to \infty} \langle v_j(y), \langle u(x), \phi(x+y) \rangle\rangle \\ &= \langle v(y), \langle u(x), \phi(x+y) \rangle\rangle = \langle u * v, \phi \rangle. \end{aligned}$$

(ii) Choose a cut-off function $\rho \in C_c^\infty(\mathbf{R}^n)$, equal to unity on a neighbourhood of K. Then one has, again from (5.1.3), that, if $\phi \in C_c^\infty(\mathbf{R}^n)$ then

$$\lim_{j\to\infty} \langle u * v_j, \phi\rangle = \lim_{j\to\infty} \langle v_j(y), \rho(y)\langle u(x), \phi(x+y)\rangle\rangle = \langle u * v, \phi\rangle$$

as $\rho(y)\langle u(x), \phi(x+y)\rangle \in C_c^\infty(\mathbf{R}^n)$, and this proves the second assertion.

We remark finally that it is clear that the convolution of a finite set of distributions can be defined in the same manner, if all but one of the distributions in the set have compact support. However, the discussion of this matter is postponed to section 5.3, where an extension of the theory will be developed.

5.2. Regularization

If $u \in \mathscr{D}'(\mathbf{R}^n)$ and $\rho \in C_c^\infty(\mathbf{R}^n)$, then one can form the convolution of u and of the distribution determined by ρ. We shall now prove that this is, in fact, a function of class C^∞. This important smoothing property of convolution generalizes Theorem 1.2.1.

Theorem 5.2.1. If $u \in \mathscr{D}'(\mathbf{R}^n)$ and $\rho \in C_c^\infty(\mathbf{R}^n)$, then

$$\rho * u(x) = \langle u(y), \rho(x-y)\rangle, \quad x \in \mathbf{R}^n, \tag{5.2.1}$$

and this function is a member of $C^\infty(\mathbf{R}^n)$.

Proof. Let $\phi \in C_c^\infty(\mathbf{R}^n)$ be given; one can regard it as a member of $\mathscr{E}'(\mathbf{R}^n)$. Take $\sigma \in C_c^\infty(\mathbf{R}^n)$ such that $\sigma = 1$ on a neighbourhood of supp ϕ; then $(x,y) \mapsto \sigma(x)\phi(x-y)$ has compact support, and so one can compute

$$\begin{aligned}&\langle \phi(x) \otimes u(y), \sigma(x)\rho(x-y)\rangle\\ &= \int \langle u(y), \rho(x-y)\rangle \phi(x)\,\mathrm{d}x = \left\langle u(y), \int \phi(x)\rho(x-y)\,\mathrm{d}x\right\rangle.\end{aligned} \tag{5.2.2}$$

Here Theorem 4.3.3 (i) has been used. Since

$$\int \phi(x)\rho(x-y)\,\mathrm{d}x = \int \phi(x+y)\rho(x)\,\mathrm{d}x,$$

another appeal to Theorem 4.3.3 shows that (5.2.2) gives

$$\int \langle u(y), \rho(x-y)\rangle \phi(x)\,\mathrm{d}x = \langle \rho(x) \otimes u(y), \phi(x+y)\rangle = \langle \rho * u, \phi\rangle$$

Finally, one has $\langle u(y), \rho(x-y)\rangle \in C^\infty(\mathbf{R}^n)$ by Theorem 4.1.1, and as ϕ can be any member of $C_c^\infty(\mathbf{R}^n)$, the theorem follows.

Definition 5.2.1. If $\rho \in C_c^\infty(\mathbf{R}^n)$ and $u \in \mathscr{D}'(\mathbf{R}^n)$, then the function $\rho * u$ is called a regularization of u.

We can now prove one of the principal results in the theory of distributions.

Theorem 5.2.2. $C_c^\infty(\mathbf{R}^n)$, considered as a subspace of $\mathscr{D}'(\mathbf{R}^n)$, is dense in $\mathscr{D}'(\mathbf{R}^n)$.

Proof. The statement means that, given any $u \in \mathscr{D}'(\mathbf{R}^n)$, one can find a sequence of test functions which converges to u in $\mathscr{D}'(\mathbf{R}^n)$. To construct such a sequence, choose some $\psi \in C_c^\infty(\mathbf{R}^n)$ such that $\int \psi \, dx = 1$, and set $\psi_j(x) = j^n \psi(jx), j = 1, 2, \ldots$. Then $\psi_j \to \delta$ in $\mathscr{D}'(\mathbf{R}^n)$ as $j \to \infty$, and $\operatorname{supp} \psi_j \subset \operatorname{supp} \psi$ for all j. Hence, by Theorem 5.1.3 (ii) and (5.1.7),

$$\psi_j * u \to \delta * u = u \quad \text{in } \mathscr{D}'(\mathbf{R}^n) \text{ as } j \to \infty. \tag{5.2.3}$$

By Theorem 5.2.1, one has $\psi_j * u \in C^\infty(\mathbf{R}^n)$. Now, take $\chi \in C_c^\infty(\mathbf{R}^n)$ such that $\chi = 1$ for $|x| < 1$, and put

$$u_j(x) = \chi(x/j)\psi_j * u(x), \quad j = 1, 2, \ldots.$$

If $\phi \in C_c^\infty(\mathbf{R}^n)$ then

$$\langle u_j, \phi \rangle = \int u_j \phi \, dx = \int (\psi_j * u)\phi \, dx$$

for all sufficiently large j, and so it follows from (5.2.3) that the u_j, which are members of $C_c^\infty(\mathbf{R}^n)$, converge to u in $\mathscr{D}'(\mathbf{R}^n)$ as $j \to \infty$. The theorem is proved.

Theorem 5.2.2 can be extended to distributions defined on an open set $X \subset \mathbf{R}^n$ by 'cutting and smoothing'. Let us first observe that one can find an increasing sequence of compact sets $K_j \subset X, j = 1, 2, \ldots$, whose union is X. For example, if $X = \mathbf{R}^n$, one can take $K_j = \{|x| \leqslant j\}$, and if $X = \{x \in \mathbf{R}^n \colon |x| < 1\}$, one can take $K_j = \{|x| \leqslant 1 - j^{-1}\}$. By adapting and combining these two examples, one can easily get a construction which works in any open set $X \subset \mathbf{R}^n$; this is left as an exercise.

Theorem 5.2.3. If $X \subset \mathbf{R}^n$ is an open set, then $C_c^\infty(X)$, considered as a subspace of $\mathscr{D}'(X)$, is dense in $\mathscr{D}'(X)$.

Proof. Let $(K_j)_{1 \leqslant j < \infty}$ be a sequence of compact subsets of X such that

$$K_j \subset K_{j+1}, \quad j = 1, 2, \ldots, \quad \bigcup_{j=1}^{\infty} K_j = X. \tag{5.2.4}$$

For each j, choose $\rho_j \in C_c^\infty(X)$ such that $\rho_j = 1$ on a neighbourhood of K_j. Let $u \in \mathscr{D}'(X)$ be given, and put $\rho_j u = u_j, j = 1, 2, \ldots$. Then $u_j \in \mathscr{E}'(X)$ extends trivially to an element of $\mathscr{E}'(\mathbf{R}^n)$. Now choose ψ such that

$$\psi \in C_c^\infty(\mathbf{R}^n), \quad \operatorname{supp} \psi \subset \{|x| \leqslant 1\}, \quad \int \psi \, dx = 1. \tag{5.2.5}$$

By Theorem 5.1.2 one can find a decreasing sequence of positive real numbers $(\epsilon_j)_{1 \leqslant j < \infty}$, tending to zero, such that, if one sets

$$\psi_j(x) = \epsilon_j^{-n} \psi(x(\epsilon_j), \quad x \in \mathbf{R}^n, \quad j = 1, 2, \ldots \tag{5.2.6}$$

then the regularizations $\psi_j * u_j$ are supported in X, and hence are elements of $C_c^\infty(X)$.

We now show that $\psi_j * u_j \to u$ as $j \to \infty$, in $\mathscr{D}'(X)$, and this will prove the theorem. If $\phi \in C_c^\infty(X)$ then, by (5.2.4), (5.2.5) and the fact that the ϵ_j tend to zero, there is a k such that $\langle u, \phi \rangle = \langle u_k, \phi \rangle$, and

$$\langle \psi_j * u_j, \phi \rangle = \left\langle u_j(x), \int \psi_j(y)\phi(x+y)\,\mathrm{d}y \right\rangle$$

$$= \left\langle u_k(x), \int \psi_j(y)\phi(x+y)\,\mathrm{d}y \right\rangle = \langle \psi_j * u_k, \phi \rangle,$$

if $j \geqslant k$. It is clear from (5.2.4) and (5.2.6) that $\psi_j \to \delta$ in $\mathscr{D}'(\mathbf{R}^n)$ as $j \to \infty$, and that $\operatorname{supp} \psi_j \subset \operatorname{supp} \psi_1$ for $j > 1$. Hence Theorem 5.1.3 and (5.1.7) again give for $j \geqslant k$,

$$\langle \psi_j * u_j, \phi \rangle = \langle \psi_j * u_k, \phi \rangle \to \langle u_k, \phi \rangle = \langle u, \phi \rangle$$

and so we are done.

5.3. Convolution of distributions with non-compact supports

So far, it has been assumed that at least one of the distributions u and v has compact support, in order to ensure the existence of the convolution $u * v$. This can be replaced by a condition that is more symmetric, and extends to any finite set of distributions. We first need a definition. Let $m > 1$ be an integer, and let

$$\mu: \mathbf{R}^{mn} \cong \mathbf{R}^n \times \ldots \times \mathbf{R}^n \to \mathbf{R}^n$$

be the map defined by

$$\mu(x^{(1)}, \ldots, x^{(m)}) = x^{(1)} + \ldots + x^{(m)}. \tag{5.3.1}$$

Definition 5.3.1. Let $A_1, \ldots, A_m$ be closed sets in $\mathbf{R}^n$. We shall say that the restriction of the map (5.3.1) to $A_1 \times \ldots \times A_m$ is proper if, for any $\delta > 0$, there is a $\delta' > 0$ such that $x^{(j)} \in A_j, j = 1, \ldots, m$ and $|x^{(1)} + \ldots + x^{(m)}| \leqslant \delta$ imply that $|x^{(j)}| \leqslant \delta'$ for $j = 1, \ldots, m$.

Note. This means that the inverse image of a compact subset of $\mathbf{R}^n$ under the map (5.3.1) is compact.

Lemma 5.3.1. Let $A_1, \ldots, A_m$ be closed subsets of $\mathbf{R}^n$, let $\epsilon > 0$, and let A_j^ϵ, $j = 1, \ldots, m$ be the closed ϵ-neighbourhoods of the A_j. Assume that the restriction of (5.3.1) to $A_1 \times \ldots \times A_m$ is proper. Then its restriction to $A_1^\epsilon, \ldots, A_m^\epsilon$ is also proper.

Proof. This will be carried out for $m = 2$; the argument in the general case is similar. We write A, B, x, and y for $A_1, A_2, x^{(1)}$ and $x^{(2)}$, respectively. Let $\delta > 0$ be given, and suppose that $x \in A, y \in B$, and $|x+y| \leqslant \delta$. There are points $x' \in A$ and $y' \in B$ such that $|x-x'| \leqslant \epsilon$ and $|y-y'| \leqslant \epsilon$. Hence

$$|x'+y'| \leqslant |x+y| + |x'-x| + |y'-y| \leqslant \delta + 2\epsilon.$$

By hypothesis, there is a $\delta' > 0$, depending only on $\delta + 2\epsilon$, such that $|x'| \leqslant \delta'$ and $|y'| \leqslant \delta'$. So

$$|x| = |x-x'| \leqslant \delta' + \epsilon, \quad |y| = |y-y'+y'| \leqslant \delta' + \epsilon,$$

and as x and y can be any points in A and B, respectively, the lemma is proved.

Let $u_1, \ldots, u_m$ be distributions on $\mathbf{R}^n$, and suppose that the restriction of the map (5.3.1) to $\operatorname{supp} u_1 \times \ldots \times \operatorname{supp} u_m$ is proper. Let $\varepsilon > 0$ and $\phi \in C_c^\infty(\mathbf{R}^n)$. By Lemma 5.3.1, the set

$$K_\varepsilon(\phi) = ((\operatorname{supp} u_1)^\varepsilon \times \ldots \times (\operatorname{supp} u_m)^\varepsilon) \cap \operatorname{supp} \phi(x^{(1)} + \ldots + x^{(m)})$$

is compact. Clearly, one can choose functions $\rho_1, \ldots, \rho_m$ in $C_c^\infty(\mathbf{R})$ such that $\rho(x^{(1)}, \ldots, x^{(m)}) = \rho_1(x^{(1)}) \otimes \ldots \otimes \rho_m(x^{(m)})$ is supported in $K_\varepsilon(\phi)$, and $\rho = 1$ on a neighbourhood of $K_0(\phi)$. Then the linear form

$$\phi \mapsto \langle u_1 \otimes \ldots \otimes u_m, \rho(x^{(1)}, \ldots, x^{(m)})\, \phi(x^{(1)} + \ldots x^{(m)})\rangle \tag{5.3.2}$$

is well defined; in effect, it is the convolution of $\rho_1 u_1, \ldots, \rho_m u_m$, which are distributions with compact supports, paired with ϕ. It is easy to check that this is independent of the choice of cut-off functions, and that it is a distribution; the details are left to the reader. It is also evident that one can omit ρ_2 when $m = 2$, and thus recover (5.1.2). So we define the convolution of $u_1, \ldots, u_m$ by (5.3.2), denote it $u_1 * \ldots * u_m$, and observe that we have proved

Theorem 5.3.1. If $u_1, \ldots, u_m \in \mathscr{D}'(\mathbf{R}^n)$, and the restriction of the map μ defined by (5.3.1) to $\operatorname{supp} u_1 \times \ldots \times \operatorname{supp} u_m$ is proper, then the convolution $u_1 * \ldots * u_m$ can be defined by (5.3.2). It reduces to the convolution $u_1 * u_2$ as defined in Theorem 5.1.1 when $m = 2$ and $u_1 \in \mathscr{E}'(\mathbf{R}^n)$.

We now list the basic properties of this extended version of convolution as a theorem.

Theorem 5.3.2. Let $u_1, \ldots, u_m$ be as in Theorem 5.3.1. Then:

(i) One has, for any $\phi \in C_c^\infty(\mathbf{R}^n)$,

$$\langle u_1 * \ldots * u_m, \phi\rangle = \langle u_1, \phi_{m-1}\rangle$$

where ϕ_{m-1} is defined recursively by

$$\phi_0 = \phi, \quad \phi_{j+1}(x) = \langle u_{m-j}(y), \phi_j(x+y)\rangle, \quad j = 0, 1, \ldots, m-2.$$

(ii) Convolution is associative; if I and J are disjoint subsets of $1, \ldots, m$ whose union is $\{1, \ldots, m\}$, then

$$u_1 * \ldots * u_m = (*u_i)_{i \in J} * (*u_j)_{j \in J}.$$

(iii) One has

$$\operatorname{supp}(u_1 * \ldots * u_m) \subset \operatorname{supp} u_1 + \ldots + \operatorname{supp} u_m$$

(iv) Let $A_1, \ldots, A_m$ be closed sets in $\mathbf{R}^n$, such that the restriction of the map (5.3.1) to $A_1 \times \ldots \times A_m$ is proper. Let $u_2, \ldots, u_m$ be distributions supported in $A_2, \ldots, A_m$, respectively, and let $(v_j)_{1 \leqslant j < \infty}$ be a sequence of distributions, all supported in A_1, which converges to a distribution u_1 in $\mathscr{D}'(\mathbf{R}^n)$. Then

$$v_j * u_2 * \ldots * u_m \to u_1 * u_2 * \ldots * u_m$$

in $\mathscr{D}'(\mathbf{R}^n)$, as $j \to \infty$.

The proofs are all immediate consequences of (5.3.2), or simple variants of the corresponding proofs in subsection 5.1. They are left to the reader, as are the extension of (5.1.5) and (5.1.6) to the present case.

The associative law can fail if the hypothesis on the supports of the 'factors' does not hold. Take $n=1$, and $u_1 = 1$ (the constant function), $u_2 = \partial\delta$, and $u_3 = H$, the Heaviside function. Then

$$(u_1 * u_2) * u_3 = (1 * \partial\delta) * H = 0 * H = 0,$$

$$u_1 * (u_2 * u_3) = 1 * (\partial\delta * H) = 1 * \delta = 1,$$

since $\partial\delta * H = \partial H = \delta$.

As an example, we consider the subspace of $\mathscr{D}'(\mathbf{R})$ consisting of distributions whose supports are 'bounded on the left':

$$\mathscr{D}'^{+}(\mathbf{R}) = \{u \in \mathscr{D}'(\mathbf{R}) \colon \operatorname{supp} u \subset [a, \infty) \text{ for some } a \in \mathbf{R}\}. \tag{5.3.3}$$

Let $a_1, \ldots, a_m$ be real numbers, and put $A_j = [a_j, \infty), j = 1, \ldots, m$. Take $\delta \geqslant 0$, and consider the set

$$\{x \in \mathbf{R}^m \colon x_j \geqslant a_j, \quad j = 1, \ldots, m, \quad |x_1 + \ldots + x_m| \leqslant \delta\}.$$

This is empty if $\delta < a_1 + \ldots + a_m$, and if $\delta \geqslant a_1 + \ldots + a_m$ one has

$$a_j \leqslant x_j \leqslant \delta + a_j - \sum_{i=1}^{m} a_i, \quad j = 1, \ldots, m.$$

So the map (5.3.1) restricted to $A_1 \times \ldots \times A_m$ is proper, and Theorem 5.3.1. applies. One concludes that the convolution of any finite set of members of $\mathscr{D}'^{+}(\mathbf{R})$ exists and that by (iii) of Theorem 5.3.2, its support is again bounded to the left. So $\mathscr{D}'^{+}(\mathbf{R})$ becomes a commutative ring if multiplication is defined to be convolution; the Dirac distribution is the unit element.

There are similar results for distributions on $\mathbf{R}$ whose supports are bounded on the right,

$$\mathscr{D}'^{-}(\mathbf{R}) = \{u \in \mathscr{D}'(\mathbf{R}) \colon \operatorname{supp} u \subset (-\infty, a] \text{ for some } a \in \mathbf{R}\}. \tag{5.3.4}$$

Using these facts, one can give an alternative construction of the primitives of a distribution $v \in \mathscr{D}'(\mathbf{R})$. Let $\rho \in C^\infty(\mathbf{R})$ be such that $\rho = 0$ when $x < a$ and $\rho = 1$ when $x > b$, when a and b are real numbers and $a < b$. Put

$$v^+ = \rho v, \quad v^- = (1 - \rho)v. \tag{5.3.5}$$

Then the support of v^+ is bounded on the left, and that of v^- is bounded on the right; the same is the case for the Heaviside function H and for $H-1$, respectively, and $\partial H = \delta$. Hence

$$u = H * v^+ + (H-1) * v^- \in \mathscr{D}'(\mathbf{R}) \tag{5.3.6}$$

is well defined, and $\partial u = v$.

The same method can be used to solve the equation

$$\partial_n u = v, \quad v \in \mathscr{D}'(\mathbf{R}^n). \tag{5.3.7}$$

Write $x \in \mathbf{R}^n$ as $x = (x', x_n)$, and put

$$E^+ = \delta(x') \otimes H(x_n), \quad E^- = \delta(x') \otimes (H(x_n) - 1). \tag{5.3.8}$$

Then

$$\partial E^+ = \partial E^- = \delta(x') \otimes \delta(x_n) = \delta(x). \tag{5.3.9}$$

It is easy to deduce from Theorem 5.3.1 that $E^+ * v^+$ is well defined if v^+ is supported in a half space $\{x_n \geqslant a\}$, where $a \in \mathbf{R}$. Similarly, $E^- * v^-$ is well defined if the support of v^- is a subset of some half space $\{x_n \leqslant b\}$, with $b \in \mathbf{R}$. Choosing $\rho \in C^\infty(\mathbf{R})$ as above, one can decompose v in this way, with

$$v^+ = v\rho(x_n), \quad v^- = v(1 - \rho(x_n)).$$

Then

$$u = E^+ * v^+ + E^- * v^- \tag{5.3.10}$$

is well defined, and satisfies (5.3.7).

We conclude this section with some remarks on convolution equations. With a distribution $k \in \mathscr{D}'(\mathbf{R}^n)$ one can associate the map $K: \mathscr{D}'(\mathbf{R}^n) \to \mathscr{D}'(\mathbf{R}^n)$ given by $u \mapsto k * u$. As the domain $D(K)$ of this one can, for example, take the set of all $u \in \mathscr{D}'(\mathbf{R}^n)$ such that the restriction of $(x, y) \mapsto x + y$ to $\operatorname{supp} k \times \operatorname{supp} u$ is a proper map; when k has compact support, this is all of $\mathscr{D}'(\mathbf{R}^n)$. A characteristic feature of the map K is that it commutes with translation, by (5.1.6).

To solve the convolution equation

$$k * u = v \tag{5.3.11}$$

for given $v \in \mathscr{D}'(\mathbf{R}^n)$ thus amounts to inverting the map K. A fundamental solution of (5.3.11) (or of the map K) is a distribution E such that

$$k * E = \delta, \tag{5.3.12}$$

where it is to be understood that $E \in D(K)$. If one has such a fundamental solution, then one can obtain a solution of (5.3.11) if v has compact support. Indeed, let

$$u = E * v. \tag{5.3.13}$$

It is easy to see that, if $E \in D(K)$ and $v \in \mathscr{E}'(\mathbf{R}^n)$, then E, v and k satisfy the hypothesis of Theorem 5.3.1. So it follows from (5.1.7), (5.3.12), and the associative law, Theorem 5.3.2 (iii), that

$$k * u = (k * E) * v = \delta * v = v. \tag{5.3.14}$$

A differential operator with constant coefficients,

$$P(\partial) = \sum_{|\alpha| \leqslant m} a_\alpha \partial^\alpha$$

defined on $\mathbf{R}^n$, can be regarded as a convolution operator. For if one takes $k = P\delta$, then, by (5.1.5) and (5.1.7),

$$P\delta * u = \delta * Pu = Pu.$$

So a fundamental solution of P is, by definition, a distribution $E \in \mathscr{D}'(\mathbf{R}^n)$ such that

$$P(\partial)E = \delta. \tag{5.3.15}$$

The adjoint (2.6.3) of P is

$${}^tP(\partial) = \sum_{|\alpha| \leqslant m} (-1)^{|\alpha|} a_\alpha \partial^\alpha = P(-\partial).$$

So (5.3.15) means that

$$\langle E, P(-\partial)\phi \rangle = \phi(0), \quad \phi \in C_c^\infty(\mathbf{R}^n). \tag{5.3.16}$$

A fundamental solution of a differential operator always gives an inverse of P on $\mathscr{E}'(\mathbf{R}^n)$, by virtue of (5.3.13) and (5.3.14). It may be possible to extend its domain; for example, E^+ given by (5.3.8) is a fundamental solution of $P = \partial_n$ on the subspace of $\mathscr{D}'(\mathbf{R}^n)$ consisting of distributions supported in a half space $\{x: x_n \geqslant a, a \in \mathbf{R}\}$.

We note finally that the difference of two fundamental solutions of a differential operator P is a (distribution) solution of the homogeneous equation $Pu = 0$.

5.4. Fundamental solutions of some differential operators

The trivial fact that the Heaviside function H is a fundamental solution of the basic differential operator ∂ on $\mathbf{R}$ can easily be extended, as follows. Write, as in (2.3.1),

$$x_+ = xH(x), \quad x \in \mathbf{R}, \tag{5.4.1}$$

so that $x_+ = x$ if $x > 0$ and $x_+ = 0$ if $x \leqslant 0$. Then $\partial x_+ = H$ and so

$$\partial^k (x_+^{k-1}/(k-1)!) = \delta, \quad k = 1, 2, \ldots.$$

It follows from this and Theorem 4.3.1 (iii) that, if one sets

$$E_k = \frac{(x_1)_+^{k-1} \ldots (x_n)_+^{k-1}}{((k-1)!)^n}, \quad x \in \mathbf{R}^n \tag{5.4.2}$$

then one has, in $\mathscr{D}'(\mathbf{R}^n)$, for $k=1,2,\ldots,$

$$(\partial_1 \ldots \partial_n)^k E_k = \delta. \tag{5.4.3}$$

This can be used to prove the so-called *structure theorem*:

Theorem 5.4.1. The restriction of a distribution $u\in\mathscr{D}'(\mathbf{R}^n)$ to a bounded open set $X\subset\mathbf{R}^n$ is a derivative of finite order of a continuous function.

Proof. As X is bounded, one can find a cut-off function $\psi\in C_c^\infty(\mathbf{R}^n)$ such that $\psi=1$ on X; then $u=\psi u$ on X. As ψu has compact support, it is of finite order. Let N be its order. By (5.4.3) one has

$$\psi u = (\partial_1 \ldots \partial_n)^{N+2} E_{N+2} * \psi u. \tag{5.4.4}$$

So the theorem will follow once it is shown that $E_{N+2} * \psi u$ is equal to a continuous function.

Take $\rho\in C_c^\infty(\mathbf{R}^n)$ as in Theorem 1.2.1,

$$\rho \geqslant 0, \quad \operatorname{supp}\rho \subset \{|x| \leqslant 1\}, \quad \int \rho \, dx = 1,$$

and set $\rho_\epsilon(x)=\epsilon^{-n}(x/\epsilon)$, where $\epsilon>0$. Consider the regularization

$$f_\epsilon = (E_{N+2} * \psi u) * \rho_\epsilon.$$

As both ψu and ρ_ϵ have compact support, the associative law applies, and gives, by an obvious variant of Theorem 5.2.1,

$$\begin{aligned} f_\epsilon &= \psi u * (E_{N+2} * \rho_\epsilon) \\ &= \langle \psi u(y), E_{N+2} * \rho_\epsilon(x-y)\rangle \end{aligned} \tag{5.4.5}$$

The proof of Theorem 1.2.1 shows that $E_{N+2} * \rho_\epsilon(z)$ converges to $E_{N+2}(z)$ in $C^N(\mathbf{R}^n)$ as $\epsilon\to 0$. As ψu is of order N, semi-norm estimates applied to (5.4.5) therefore show that the continuous functions f_ϵ converge, uniformly when x is in a compact set, to $f(x)=\langle u(y), E_{N+2}(x-y)\rangle$ as $\epsilon\to 0$. Hence this is also a continuous function. On the other hand, f_ϵ is a regularization which converges to $(\psi u) * E_{N+2}$ in $\mathscr{D}'(\mathbf{R}^n)$ as $\epsilon\to 0$. It is clear that the limits are the same in this case, so that

$$E_{N+2} * \psi u(x) = f(x) = \langle \psi u(y), E_{N+2}(x-y)\rangle \in C^0(\mathbf{R}^n).$$

But $u=\psi u$ on X, so the theorem now follows from (5.4.4).

Note. One can of course add a function to f which is equal to a polynomial of degree $\leqslant N+1$ on X. Note also that the support of f can be in any pre-assigned neighbourhood of the closure of X.

When u has compact support, then $u=\sigma u$ if $\sigma=1$ on a neighbourhood of the support of u. Furthermore, one can take $\psi=1$ in the proof of Theorem 5.4.1. So one has

$$\langle u, \phi \rangle = \langle u, \sigma\phi \rangle = (-1)^{(N+1)n} \int f(\partial_1 \ldots \partial_n)^{N+1} \sigma\phi \, dx, \quad \phi \in C_c^\infty(\mathbf{R}^n),$$

where N is a nonnegative integer and f is a continuous function. Expanding by Leibniz's theorem, one therefore has:

Corollary 5.4.1. Let $u \in \mathscr{E}'(\mathbf{R}^n)$, then there is an integer $m \geqslant 0$ and a set of continuous functions f_α, $|\alpha| \leqslant m$, such that

$$u = \sum_{|\alpha| \leqslant m} \partial^\alpha f_\alpha.$$

We shall now obtain fundamental solutions for some classical differential operators. We begin with the *Laplacian*,

$$\Delta = \partial_1^2 + \ldots + \partial_n^2. \tag{5.4.6}$$

As it is invariant under the rotation group, it is natural to look for a fundamental solution that is also invariant under this group. In polar coordinates $r \in \mathbf{R}^+$, $\theta \in \mathbf{S}^{n-1}$, such that $x = r\theta$, the Laplacian, acting on C^∞ functions, is

$$\Delta = \partial_r^2 + \frac{h-1}{r}\, \partial_r + \frac{1}{r^2}\, \Delta_S$$

where Δ_S is the Laplace-Beltrami operator on $\mathbf{S}^{n-1}$. So, if $n > 2$, then the only classical solutions of Laplace's equation on $\mathbf{R}^n \setminus \{0\}$ that are functions of $r = |x|$ only are of the form $A + Br^{2-n}$, where A and B are constants. When $n = 2$, this must be replaced by $A + B \log r$.

Assume that $n > 2$. Since $x \mapsto |x|^{2-n}$ is locally integrable, it determines a distribution. Set

$$\Delta |x|^{2-n} = u,$$

where $u \in \mathscr{D}'(\mathbf{R}^n)$. Then $u = 0$ on $\mathbf{R}^n \setminus \{0\}$, that is to say $\operatorname{supp} u = \{0\}$. Hence

$$u = \sum_{\alpha \geqslant 0} c_\alpha \partial^\alpha \delta,$$

by Theorem 3.2.1, where the c_α are complex numbers, and the sum is finite. Now $x \mapsto |x|^{2-n}$ is homogeneous of degree $2-n$; so u is homogeneous of degree $-n$ (Exercise 4.3). By (4.2.12), one therefore has

$$\sum_{\alpha \geqslant 0} c_\alpha \partial^\alpha \delta - \sum_{\alpha \geqslant 0} t^{-|\alpha|} c_\alpha \partial^\alpha \delta = 0, \quad t > 0.$$

By testing this against $\phi x^\alpha / \alpha!$, where $\phi \in C_c^\infty(\mathbf{R}^n)$ and $\phi = 1$ on a neighbourhood of the origin, one concludes that $c_\alpha = 0$ for $\alpha > 0$. So

$$\Delta |x|^{2-n} = c\delta,$$

where c is a constant. To determine this, take $\psi(t)\in C_c^\infty(\mathbf{R})$ such that $\psi=1$ on a neighbourhood of $t=0$. Then $\psi(|x|)\in C_c^\infty(\mathbf{R}^n)$, and so

$$c=\langle\Delta|x|^{2-n},\psi(|x|)\rangle=\langle|x|^{2-n},\Delta\psi(|x|)\rangle$$

$$=\omega_{n-1}\int_0^\infty r^{2-n}\left(\partial^2\psi(r)+\frac{n-1}{r}\partial\psi(r)\right)r^{n-1}\,dr=(2-n)\omega_{n-1},$$

where $\omega_{n-1}=2\pi^{1/2n}/\Gamma(\frac{1}{2}n)$ is the area of $\mathbf{S}^{n-1}$.

We have thus proved that the distribution determined by the locally integrable function

$$E=1/(n-2)\omega_{n-1}|x|^{n-2} \tag{5.4.7}$$

is a fundamental solution of $-\Delta$ on $\mathbf{R}^n$ if $n>2$. For $n=3$, this becomes the well known $E=1/4\pi|x|$.

When $n=2$, this has to be replaced by

$$E=-\frac{1}{2\pi}\log|x|\in\mathscr{D}'(\mathbf{R}^2). \tag{5.4.8}$$

This is elementary (the logarithmic potential) and can be proved directly, or in the same way as (5.4.7).

Next, let $x\in\mathbf{R}^2$, and put

$$z=x_1+\mathrm{i}x_2,\bar z=x_1-\mathrm{i}x_2.$$

Then, if $f\in C^1(\mathbf{R}^2)$, one has

$$\mathrm{d}f=\frac{\partial f}{\partial z}\,\mathrm{d}z+\frac{\partial f}{\partial\bar z}\,\mathrm{d}\bar z,$$

where

$$\frac{\partial}{\partial z}=\tfrac{1}{2}(\partial_1-\mathrm{i}\partial_2),\quad \frac{\partial}{\partial\bar z}=\tfrac{1}{2}(\partial_1+\mathrm{i}\partial_2). \tag{5.4.9}$$

The differential operator $\partial/\partial\bar z$ is called the *Cauchy-Riemann* operator, as the Cauchy-Riemann equations can be written jointly as $\partial f/\partial\bar z=0$. In particular, one has $(\partial/\partial\bar z)(1/z)=0$ on $\mathbf{R}^2\setminus\{0\}$. As $x\mapsto 1/z$ is locally integrable, it determines a distribution, which we also denote by $1/z$. We shall now show that

$$\frac{\partial}{\partial\bar z}\frac{1}{z}=\pi\delta. \tag{5.4.10}$$

If $\phi\in C_c^\infty(\mathbf{R}^2)$, then

$$\left\langle\frac{\partial}{\partial\bar z}\frac{1}{z},\phi\right\rangle=-\left\langle\int\frac{1}{z},\frac{\partial\phi}{\partial\bar z}\right\rangle=-\int\frac{1}{z}\frac{\partial\phi}{\partial\bar z}\,\mathrm{d}x$$

$$=-\lim_{\epsilon\to0+}\int_{|x|>\epsilon}\frac{\partial}{\partial\bar z}\left(\frac{\phi}{z}\right)\mathrm{d}x=\lim_{\epsilon\to0+}\int_{|x|=\epsilon}\frac{\phi}{z}(\mathrm{d}x_2-\mathrm{i}\mathrm{d}x_1),$$

where the last expression is obtained by Stokes's theorem, the integral being over the circle $\{|x|=\epsilon\}$, with counterclockwise orientation. So

$$\frac{\partial}{\partial \bar{z}}\left\langle \frac{1}{z}, \phi \right\rangle = \lim_{\epsilon \to 0+} \tfrac{1}{2} \int_0^{2\pi} \phi(\epsilon \cos\theta, \epsilon \sin\theta)\, d\theta = \phi(0), \quad \phi \in C_c^\infty(\mathbf{R}^2),$$

which is (5.4.10).

Finally, we consider the *heat operator*. For this, we introduce an additional real variable t, and work in $\mathbf{R}^{n+1} \cong \mathbf{R}^n \times \mathbf{R}$, with a generic point written as (x, t). The heat operator is then

$$P = \partial_t - \Delta. \tag{5.4.11}$$

It is well known, and easy to check, that the locally integrable function

$$E(x, t) = H(t) 2^{-n} (\pi t)^{-1/2n} \exp(-|x|^2/4t) \tag{5.4.12}$$

satisfies $PE = 0$ on $\mathbf{R}^n \times (\mathbf{R} \backslash \{0\})$. We shall show that it is a fundamental solution of P,

$$PE = \delta. \tag{5.4.13}$$

Indeed, if $\phi \in C_c^\infty(\mathbf{R}^n \times \mathbf{R})$, then

$$\begin{aligned}\langle PE, \phi\rangle = \langle E, {}^tP\phi\rangle &= -\int E(x,t)(\partial_t\phi + \Delta\phi)\, dx\, dt \\ &= -\lim_{\epsilon\to 0+} \int_{t>\epsilon} E(x,t)(\partial_t\phi + \Delta\phi)\, dx\, dt \\ &= \lim_{\epsilon\to 0+} \int E(x,\epsilon)\phi(x,\epsilon)\, dx,\end{aligned}$$

since $E \in C^\infty(\mathbf{R}^n \times \mathbf{R}^+)$ and $PE = 0$ for $t > 0$. Put $x = 2\epsilon^{1/2} y$, to obtain

$$\langle PE, \phi\rangle = \pi^{-\frac{1}{2}n} \lim_{\epsilon\to 0+} \int \phi(2\epsilon^{1/2} y, \epsilon) \exp(-|y|^2)\, dy.$$

As ϕ is bounded, and $\exp(-|y|^2) \in L_1(\mathbf{R}^n)$, one can make $\epsilon \to 0$ under the integral sign, by dominated convergence. This gives

$$\langle PE, \phi\rangle = \pi^{-\frac{1}{2}n} \int \phi(0,0) \exp(-|y|^2)\, dy = \phi(0,0), \quad \phi \in C_c^\infty(\mathbf{R}^n \times \mathbf{R}),$$

which is (5.4.13).

One can use the fundamental solution (5.4.12) to derive the solution of a generalized initial value problem. The classical initial value problem can be formulated as

$$\begin{aligned}&Pv = 0 \quad \text{on} \quad \mathbf{R}^n \times \mathbf{R}^+, \\ &v \in C^2(\mathbf{R}^n \times \mathbf{R}^+) \cap C^0(\mathbf{R}^n \cap \bar{\mathbf{R}}^+), \quad v(x,0) = f(x),\end{aligned} \tag{5.4.14}$$

where

$$C^0(\mathbf{R}^n \times \bar{\mathbf{R}}^+) = \{v \in C^0(\mathbf{R}^n \times \mathbf{R}^+): v = \tilde{v} | \mathbf{R}^n \times \mathbf{R}^+ \quad \text{for some} \quad \tilde{v} \in C^0(\mathbf{R}^n \times \mathbf{R})\}.$$

An equivalent condition is that v should extend by continuity to a continuous function on $\mathbf{R}^n \times \bar{\mathbf{R}}^+$; one can then define $\tilde{v}$, for example, by setting $\tilde{v}(x,t) = v(x,-t)$ for $t<0$.

There is a device for converting (5.4.14) into a differential equation for a distribution which is important in the treatment of boundary value problems. Suppose that the problem has a solution v, and set

$$v^c = v, \quad t>0, \quad v^c = 0, \quad t \leqslant 0. \tag{5.4.15}$$

Thus v^c is obtained by 'cutting off' an extension of v to $\mathbf{R}^n \times \mathbf{R}$ at the boundary $\mathbf{R}^n \times \{0\}$ of $\mathbf{R}^n \times \mathbf{R}^+$. To compute Pv^c, one has

$$\begin{aligned}\langle Pv^c, \phi \rangle = \langle v^c, {}^tP\phi \rangle &= -\int_{t>0} v(\partial_t + \Delta)\phi \, \mathrm{d}x \, \mathrm{d}t \\ &= \int f(x)\phi(x,0) \, \mathrm{d}x, \quad \phi \in C_c^\infty(\mathbf{R}^n \times \mathbf{R}),\end{aligned}$$

by an integration by parts, and (5.4.14). Hence

$$Pv^c = f(x) \otimes \delta(t). \tag{5.4.16}$$

This, supplemented by the condition supp $v^c \subset \mathbf{R}^n \times \bar{\mathbf{R}}^+$, is the 'distributional' form of (5.4.14).

These considerations suggest the following generalized initial value problem: to find $u \in \mathscr{D}'(\mathbf{R}^n \times \mathbf{R})$ such that

$$Pu = f(x) \otimes \delta(t) \tag{5.4.17}$$

where $f \in \mathscr{D}'(\mathbf{R}^n)$, and that

$$\operatorname{supp} u \subset \mathbf{R}^n \times \bar{\mathbf{R}}^+. \tag{5.4.18}$$

Let us take $f \in \mathscr{E}'(\mathbf{R}^n)$. Then $f \otimes \delta$ has compact support, so that

$$u = E(x,t) * (f(x) \otimes \delta(t)) \tag{5.4.19}$$

satisfies (5.4.17). Again, (5.4.18) follows from Theorem 5.1.2, since the support of E is $\mathbf{R}^n \times \bar{\mathbf{R}}^+$, and the support of $f \otimes \delta$ is a subset of $\mathbf{R}^n \times \{0\}$.

We shall now show that the restriction of (5.4.19) to $\mathbf{R}^n \times \mathbf{R}^+$ is a C^∞ function which converges to f in $\mathscr{D}'(\mathbf{R}^n)$ as $t \to 0+$. For any $\phi \in C_c^\infty(\mathbf{R}^n \times \mathbf{R})$, one has

$$\begin{aligned}\langle u, \phi \rangle &= \langle E(x,t) \otimes f(y) \otimes \delta(s), \phi(x+y, s+t) \rangle \\ &= \left\langle f(y), \int E(x,t)\phi(x+y,t) \, \mathrm{d}x \, \mathrm{d}t \right\rangle \\ &= \left\langle f(y), \int E(x-y,t)\phi(x,t) \, \mathrm{d}x \, \mathrm{d}t \right\rangle.\end{aligned}$$

Now E, restricted to $\mathbf{R}^n \times \mathbf{R}^+$, is C^∞; hence one can appeal to Theorem 4.3.3 (i) to conclude that, if $\phi \in C_c^\infty(\mathbf{R}^n \times \mathbf{R}^+)$, then

$$\left\langle f(y), \int E(x-y,t)\phi(x,t)\,\mathrm{d}x\,\mathrm{d}t \right\rangle$$
$$= \langle f(y) \otimes \phi(x,t), \sigma(x,y,t)E(x-y,t)\rangle$$
$$= \int \langle f(y), E(x-y,t)\rangle \phi(x,t)\,\mathrm{d}x\,\mathrm{d}t.$$

(Here, $\sigma \in C_c^\infty(\mathbf{R}^n \times \mathbf{R}^n \times \mathbf{R}^+)$ is such that $\sigma = 1$ on a neighbourhood of supp $f \times$ supp ϕ.) Hence

$$u = \langle f(y), E(x-y,t)\rangle, \quad t > 0. \tag{5.4.20}$$

By Corollary 4.1.2, this is C^∞ for $t > 0$, and so satisfies $Pu = 0$ then in the usual sense.

Again, if $\phi \in C_c^\infty(\mathbf{R}^n)$, then one has for $t > 0$

$$\int E(x,t)\phi(x)\,\mathrm{d}x = \frac{1}{2^n(\pi t)^{\frac{1}{2}n}} \int \phi(x) \exp(-|x|^2/4t)\,\mathrm{d}x$$
$$= \pi^{-\frac{1}{2}n} \int \phi(2yt^{1/2}) \exp(-|y|^2)\,\mathrm{d}y \to \phi(0,0)$$

as $t \to 0+$, by dominated convergence. Thus

$$\lim_{t \to 0+} E(x,t) = \delta(x) \tag{5.4.21}$$

in $\mathscr{D}'(\mathbf{R}^n)$. So it follows from (5.4.20) and Theorems 5.2.1 and 5.1.3 (i) that $u \to f$ in $\mathscr{D}'(\mathbf{R}^n)$ as $t \to 0+$.

Exercises

5.1. (i) Give an example of two closed sets $A \subset \mathbf{R}$ and $B \subset \mathbf{R}$ whose vector sum $A+B$ is not closed.

(ii) Prove that, if A and B are closed sets in $\mathbf{R}^n$, and the restriction of the map $(x,y) \mapsto x+y$ to $A+B$ is proper, then $A+B$ is closed.

5.2. Let $u \in \mathscr{E}'(\mathbf{R})$, and put

$$U(z) = \frac{1}{2\pi \mathrm{i}} \left\langle u(t), \frac{1}{t-z} \right\rangle,$$

where $z = x + \mathrm{i}y \in \mathbf{C}$ and $y \neq 0$. Show that U is analytic on the complement of $\{z : x = \operatorname{supp} u, y = 0\}$, and that

$$U(x+\mathrm{i}y) - U(x-\mathrm{i}y) \to u \quad \text{in} \quad \mathscr{D}'(\mathbf{R}) \quad \text{as} \quad y \to 0+.$$

5.3. Let λ be a complex number, and let E_λ be the distribution (2.3.8) which is equal to $x_+^{\lambda-1}/\Gamma(\lambda) \in L_1^{\mathrm{loc}}(\mathbf{R})$ when Re $\lambda > 0$, and defined by analytic continuation in λ for all $\lambda \in \mathbf{C}$. Prove that

$$E_\lambda * E_\mu = E_{\lambda+\mu}.$$

Deduce that, if $\lambda \notin Z$, and $v \in \mathscr{D}'(\mathbf{R})$, then the convolution equation

$$x_+^{\lambda-1} * u = v$$

has a unique solution in $\mathscr{D}'^{+}(\mathbf{R})$ which can be put into the form

$$u = \frac{\sin \pi\lambda}{\pi} \partial(x_+^{-\lambda} * v).$$

5.4. Show that, if $u \in \mathscr{E}'(\mathbf{R}^n)$ and $\psi \in C^\infty(\mathbf{R}^n)$, then

$$u * \psi = \langle u(y), \psi(x-y)\rangle \in C^\infty(\mathbf{R}^n).$$

Prove also that, if u is of order N and $\psi \in C^N(\mathbf{R}^n)$, then $u * \psi$ is of the same form, and is a continuous function on $\mathbf{R}^n$.

5.5. Write $x \in \mathbf{R}^n$ as (x', x_n), where $x' \in \mathbf{R}^{n-1}$, and put

$$\Gamma = \{x : x_n \geqslant c|x'|\}$$

where c is a positive real number, and let

$$\mathscr{D}'_\Gamma(\mathbf{R}^n) = \{u \in \mathscr{D}'(\mathbf{R}^n); \operatorname{supp} u \subset \Gamma\}.$$

Show that, if $u_1, \ldots, u_m$ are all in $\mathscr{D}'_\Gamma(\mathbf{R}^n)$, then the convolution $u_1 \ldots u_m$ is well defined, and is again in $\mathscr{D}'_\Gamma(\mathbf{R}^n)$. Show also that, if v is another distribution and $\operatorname{supp} v \subset \{x_n \geqslant a, a \in \mathbf{R}\}$, then $u_1 \ldots u_m * v$ is well defined.

Suppose now that the convolution operator $k \in \mathscr{D}'_\Gamma(\mathbf{R}^n)$ has a fundamental solution $E \in \mathscr{D}'_\Gamma(\mathbf{R}^n)$, and that $v \in \mathscr{D}'(\mathbf{R}^n)$ is supported in $\{x_n \geqslant 0\}$. Show that the convolution equation $k * u = v$ has one and only one solution u such that $\operatorname{supp} u \in \{x_n \geqslant 0\}$. Find an outer bound for its support.

5.6. The differential operator

$$P(\partial) = \partial_2^2 - \partial_1^2$$

is defined on $\mathbf{R}^2$. Show that

$$E = \tfrac{1}{2} \quad \text{if} \quad x_2 > |x_1|, \qquad E = 0 \quad \text{if} \quad x_2 \leqslant |x_1|$$

is a fundamental solution of P. Deduce that, if $f \in \mathscr{D}'(\mathbf{R}^2)$ and $\operatorname{supp} f \subset \{x_2 \geqslant 0\}$, then there is a unique $u \in \mathscr{D}'(\mathbf{R}^2)$ such that

$$Pu = f, \quad \operatorname{supp} u \subset \{x_2 \geqslant 0\}$$

Obtain a representation of u, and show that in the case of continuous f one has

$$u(x) = \tfrac{1}{2} \int_\Omega f(y)\, dy, \quad \text{where} \quad \Omega = \{y : y_2 < x_2 - |x_1 - y_1|\}.$$

5.7. This exercise outlines a method for constructing a member of $C_c^\infty(\mathbf{R})$ which allows some control over the magnitude of its derivatives.

Let $t_0, t_1, \ldots$ be positive real numbers such that $\Sigma_{k=0}^\infty t_k = 1$, and let $\phi_0 \in C_c^0(\mathbf{R})$ be such that

$$\phi_0 \geqslant 0, \quad \operatorname{supp} \phi_0 \subset [0, t_0], \quad \int \phi_0\, dx = 1.$$

Put $\chi_k(x) = 1/t_k$ if $0 < x < t_k$, $\chi_k(x) = 0$ for all other x, and define

$$\phi_k = \chi_k * \phi_{k-1}, \quad k = 1, 2, \ldots.$$

Also, put

$\mu_0 = \sup \phi_0, \quad \mu_k = 2^{k+1} \sup \phi_0 / t_1 \dots t_k, \quad k = 1, 2, \dots.$

Prove that:

(i) $0 \leqslant \phi_k \leqslant \mu_0, \quad \operatorname{supp} \phi_k \subset \left[0, \sum_{j=0}^{k} t_j\right], \quad \int \phi_k \, dx = 1,$

(ii) $\phi_k \in C^k(\mathbf{R})$, and $|\partial^k \phi_k| \leqslant \phi_k, \quad k = 1, 2, \dots;$

(iii) $|\partial^j \phi_k| \leqslant \mu_j$ if $1 \leqslant j < k$ and $k = 1, 2, \dots.$

(iv) Now use the identity

$$\phi_{k+1}(x) - \phi_k(x) = \frac{1}{t_{k+1}} \int_0^{t_{k+1}} (\phi_k(x-t) - \phi_k(x)) \, dt$$

to prove that the ϕ_k converge to a function $\phi \in C_c^\infty(\mathbf{R})$ which is such that

$$\phi \geqslant 0, \quad \operatorname{supp} \phi \subset [0, 1], \quad \int \phi \, dx = 1, \quad \text{and}$$

$$|\partial^k \phi| \leqslant \mu_k, \quad k = 1, 2, \dots.$$

6 DISTRIBUTION KERNELS

An integral transform is a map of the form

$$f \mapsto kf = \int_Y k(x,y)f(y)\,\mathrm{d}y$$

where k, which is called the kernel of the transform, is a function defined on the product of an open set $X \subset \mathbf{R}^n$ and an open set $Y \subset \mathbf{R}^m$. Such a transform maps a suitable class of functions defined on Y to functions on X.

One can extend this to distributions by starting with an element k of $\mathscr{D}'(X \times Y)$. This generates a sequentially continuous map $C_c^\infty(Y) \to \mathscr{D}'(X)$; conversely, the kernel theorem says that any such map is generated by a distribution kernel. Such kernels are important in the theory of linear differential equations.

6.1. Schwartz kernels and the kernel theorem

Let $X \subset \mathbf{R}^n$ and $Y \subset \mathbf{R}^m$ be open sets, and let $k \in \mathscr{D}'(X \times Y)$. Then

$$(\phi, \psi) \mapsto \langle k, \phi \otimes \psi \rangle, \quad \phi \in C_c^\infty(X), \quad \psi \in C_c^\infty(Y) \tag{6.1.1}$$

is a bilinear form on $C_c^\infty(X) \times C_c^\infty(Y)$. For fixed ψ, it becomes a linear form on $C_c^\infty(X)$, and for fixed ϕ it becomes a linear form on $C_c^\infty(Y)$. Both of these are distributions.

Indeed, if $K \subset X$ and $K' \subset Y$ are compact sets, then there is a semi-norm estimate

$$|\langle k, \phi \otimes \psi \rangle| \leqslant C \sum_{|\alpha|+|\beta| \leqslant N} \sup |\partial_x^\alpha \phi \partial_y^\beta \psi|, \quad \phi \in C_c^\infty(K), \quad \psi \in C_c^\infty(K').$$

It is helpful to replace this by the equivalent estimate

$$|\langle k, \phi \otimes \psi \rangle| \leqslant C \sum_{|\alpha| \leqslant N} \sup |\partial^\alpha \phi| \sum_{|\beta| \leqslant N} \sup |\partial^\beta \psi|, \qquad \phi \in C_c^\infty(K), \quad \psi \in C_c^\infty(K'). \tag{6.1.2}$$

Write $k\psi$ for the linear form $\phi \mapsto \langle k, \phi \otimes \psi \rangle$:

$$\langle k\psi, \phi\rangle = \langle k, \phi \otimes \psi\rangle, \quad \phi \in C_c^\infty(X). \tag{6.1.3}$$

Then (6.1.2) gives a semi-norm estimate for $k\psi$ when ψ is fixed, so that $k\psi \in \mathscr{D}'(X)$. Moreover, (6.1.2) shows that the map $\psi \mapsto k\psi$ is sequentially continuous: if $(\psi_j)_{1 \leq j < \infty}$ is a sequence converging to 0 in $C_c^\infty(Y)$ as $j \to \infty$, then $k\psi_j \to 0$ in $\mathscr{D}'(X)$. We shall say that the map is generated by k as a *distribution kernel*, or as a *Schwartz kernel*. It is customary to use the same letter for the kernel, and for the map generated by it.

It is also obvious from (6.1.2) that, for fixed ϕ, $\psi \mapsto \langle k, \phi \otimes \psi\rangle$ is a sequentially continuous map $C_c^\infty(X) \to \mathscr{D}'(Y)$. To write this in a similar way, one can first define the *transposed kernel* ${}^t k \in \mathscr{D}'(Y \times X)$ of k, as follows. The trivial isomorphism $X \times Y \to Y \times X$ given by $(x, y) \mapsto (y, x)$ induces a pullback map $C_c^\infty(Y \times X) \to C_c^\infty(X \times Y)$,

$${}^t\chi(x, y) = \chi(y, x), \quad \chi \in C_c^\infty(Y \times X) \tag{6.1.4}$$

which is obviously an isomorphism. This in turn induces an isomorphism $\mathscr{D}'(X \times Y) \to \mathscr{D}'(Y \times X)$.

$$\langle {}^t k, \chi\rangle = \langle k, {}^t\chi\rangle, \quad k \in \mathscr{D}'(X \times Y), \quad \chi \in C_c^\infty(Y \times X). \tag{6.1.5}$$

It follows from (6.1.4) and (6.1.5) that the map ${}^t k$: $C_c^\infty(X) \to \mathscr{D}'(Y)$ generated by the transpose ${}^t k$ of k is

$$\langle {}^t k\phi, \psi\rangle = \langle k, \phi \otimes \psi\rangle, \quad \psi \in C_c^\infty(Y). \tag{6.1.6}$$

This notation thus makes it possible to discuss the two maps associated with the bilinear form (6.1.1) under one head.

It is important to observe that the maps k and ${}^t k$ are adjoints of each other; in fact, one has

$$\langle k\psi, \phi\rangle = \langle {}^t k\phi, \psi\rangle = \langle k, \phi \otimes \psi\rangle, \quad \phi \in C_c^\infty(X), \quad \psi \in C_c^\infty(Y), \tag{6.1.7}$$

Furthermore, each of the maps k and ${}^t k$ determines the other one.

Let us consider two simple examples. If $k \in C_c^\infty(X \times Y)$, then (6.1.3) yields the classical integral transform,

$$k\psi(x) = \int k(x, y)\psi(y)\, \mathrm{d}y, \quad \psi \in C_c^\infty(Y)$$

and the adjoint transform is

$${}^t k\phi(y) = \int k(x, y)\phi(x)\, \mathrm{d}x, \quad \phi \in C_c^\infty(X).$$

If $k = u \otimes v$ where $u \in \mathscr{D}'(X)$ and $v \in \mathscr{D}'(Y)$, then the ranges of both maps are one-dimensional:

$$k\psi = \langle v, \psi\rangle u, \quad {}^t k\phi = \langle u, \phi\rangle v.$$

Suppose now that one is given a sequentially continuous linear map μ: $C_c^\infty(Y) \to \mathscr{D}'(X)$. Then it is natural to ask whether there is a kernel which generates this map. The main content of the *Schwartz kernel theorem* is that this is indeed the case. We shall now give a proof of this, although it is, strictly speaking, beyond the scope of this book, as it requires an appeal to a theorem of Functional Analysis.

The kernel theorem was first deduced by L. Schwartz from A. Grothendieck's theory of nuclear spaces; see [7] for a detailed account. Simpler proofs were then published by L. Ehrenpreis (*Proc. Amer. Math Soc.* **7**, 1956, 713-18) and by H. Gask (*Math. Scand.* **8**, 1960, 327-32). The proof below is based on unpublished lecture notes by L. Hörmander.

Theorem 6.1.1. Let $X \subset \mathbf{R}^n$ and $Y \subset \mathbf{R}^m$ be open sets. A linear map μ: $C_c^\infty(Y) \to \mathscr{D}'(X)$ is sequentially continuous if and only if it is generated by a Schwartz kernel $k \in \mathscr{D}'(X \times Y)$,

$$\langle \mu\psi, \phi\rangle = \langle k, \phi \otimes \psi\rangle, \quad \phi \in C_c^\infty(X), \quad \psi \in C_c^\infty(Y). \tag{6.1.8}$$

Moreover, the kernel k is uniquely determined by μ.

Note. For clarity, the map μ and the kernel k are distinguished in the enunciation. But, once the theorem has been proved, one can write k for μ again.

Proof. We have already established sufficiency. What remains to be proved, and is harder, is that, given μ, one can find a distribution kernel k such that (6.1.8) holds. Uniqueness, however, follows at once from Theorem 4.3.1, since (6.1.8) determines k on a dense subspace of $C_c^\infty(X \times Y)$. The nub of the matter is the existence proof. This will be carried out for $X = \mathbf{R}^n$ and $Y = \mathbf{R}^m$. The argument also works when X and Y are rectangles, and the general case is easily reduced to this one by a partition of unity.

Let us, then, assume that we are given a sequentially continuous linear map μ: $C_c^\infty(\mathbf{R}^m) \to \mathscr{D}'(\mathbf{R}^n)$. Define a bilinear form B on $C_c^\infty(\mathbf{R}^n) \times C_c^\infty(\mathbf{R}^m)$ by setting

$$B(\phi, \psi) = \langle \mu\psi, \phi\rangle, \quad \phi \in C_c^\infty(\mathbf{R}^n), \quad \psi \in C_c^\infty(\mathbf{R}^m). \tag{6.1.9}$$

Our first step is to derive an estimate similar to (6.1.2). Let b be a positive real number, and put

$$\begin{aligned} K_b &= \{x \in \mathbf{R}^n : |x_i| \leqslant b, \quad i = 1, \ldots, n\} \\ K'_b &= \{y \in \mathbf{R}^m : |y_j| \leqslant b, \quad j = 1, \ldots, m\} \end{aligned} \tag{6.1.10}$$

Both K_b and K'_b are compact sets. As $\mu\psi \in \mathscr{D}'(\mathbf{R}^n)$, there is a semi-norm estimate

$$|B(\phi, \psi)| \leqslant C_\psi \sum_{|\alpha| \leqslant N_\psi} \sup |\partial^\alpha \phi|, \quad \phi \in C_c^\infty(K_b),$$

where C_ψ and N_ψ depend on ψ. Again, since μ is sequentially continuous, $\psi \mapsto B(\phi, \psi)$ is an element of $\mathscr{D}'(\mathbf{R}^m)$ for each $\phi \in C_c^\infty(\mathbf{R}^n)$, and so one also has a semi-norm estimate

$$|B(\phi,\psi)|\leqslant C'_\phi \sum_{|\beta|\leqslant M_\phi} \sup|\partial^\beta\psi|, \quad \psi\in C_c^\infty(K'_b).$$

These two inequalities show that the restriction of B to $C_c^\infty(K_b)\times C_c^\infty(K'_b)$ is a separately continuous bilinear form. Both of these are Fréchet spaces, as explained in the Appendix. One can therefore appeal to a theorem of Functional Analysis which says that a separately continuous bilinear form on the product of two Fréchet spaces is also jointly continuous [4, p. 51]. In the present case this implies that there are constants $C, N\geqslant 0$ such that

$$|B(\phi,\psi)|\leqslant C\sum_{|\alpha|\leqslant N}\sup|\partial^\alpha\phi| \sum_{|\beta|\leqslant N}\sup|\partial^\beta\psi|, \tag{6.1.11}$$

Note that this is, in effect, the estimate (6.1.2), which has thus been recovered. One could take the sum of the sup $|\partial^\beta\psi|$ over $|\beta|\leqslant M\neq N$, but nothing is gained by this.

The second step in the proof is to show that (6.1.11) implies that B extends to a distribution on $X\times Y$. For this, we first work on $K_a\times K'_a$, where $0<a<b$, and K_a, K'_a are as in (6.1.10), with a instead of b. Let $\chi\in C_c^\infty(K_a\times K'_a)$, and choose cut-off functions $\rho\in C_c^\infty(K_b)$ and $\sigma\in C_c^\infty(K'_b)$ such that $\rho=1$ on K_a and $\sigma=1$ on K'_a. Then one has, by expanding the function on $\mathbf{R}^n\times\mathbf{R}^m$ which is equal to χ on $K_b\times K'_b$, and has period b in each argument,

$$\chi(x,y)=\sum_{\mathbf{Z}^n\times\mathbf{Z}^m}\hat{\chi}_{g,h}\rho(x)\sigma(y)E(g\cdot x)E(h\cdot y). \tag{6.1.12}$$

The notation here is as follows:

$$E(t)=\exp(2\pi it/b), \quad t\in\mathbf{R},$$

and

$$\hat{\chi}_{g,h}=b^{-n-m}\int_{K_b\times K'_b}\chi(x,y)E(-g\cdot x-h\cdot y)\,dx\,dy. \tag{6.1.13}$$

(Compare the proof of Lemma 4.3.1.)

It follows from (6.1.11) that there is a constant C_1 such that

$$|B(\rho(x)E(g\cdot x),\sigma(y)E(h\cdot y))|\leqslant C_1(1+|g|)^N, \quad (g,h)\in\mathbf{Z}^n\times\mathbf{Z}^m.$$

On the other hand, it is easy to deduce from (6.1.13), by integration by parts, that, for any integer $M\geqslant 0$ there is a $C_2=C_2(M)$ such that

$$(1+|g|)^M(1+|h|)^M|\hat{\chi}|\leqslant C_2\sum_{\substack{|\alpha|\leqslant M\\|\beta|\leqslant M}}\sup|\partial_x^\alpha\partial_y^\beta\chi|,$$

$$(g,h)\in\mathbf{Z}^n\times\mathbf{Z}^m. \tag{6.1.14}$$

Combining these two estimates, one obtains

$$\sum_{Z^n \times Z^m} |\hat{\chi}_{g,h} B(\rho(x)E(g\cdot x), \sigma(y)E(h\cdot y))|$$

$$\leqslant C_3 \sum_{\substack{|\alpha|\leqslant M\\|\beta|\leqslant M}} \sup |\partial_x^\alpha \partial_y^\beta \chi| \tag{6.1.15}$$

where

$$C_3 = C_1 C_2 \sum_{Z^n \times Z^m} (1+|g|)^{N-M}(1+|h|)^{N-M} < \infty,$$

if one takes M large enough.

The inequality (6.1.15) shows that one can, guided by (6.1.9) and (6.1.12), define a continuous linear form k on $C_c^\infty(K_a \times K_a')$ by

$$\langle k, \chi\rangle = \sum_{Z^n \times Z^m} \hat{\chi}_{g,h} B(\rho(x)E(g\cdot x), \sigma(y)E(h\cdot y)). \tag{6.1.16}$$

For the series converges, and the absolute value of its sum is bounded by the second member of (6.1.15).

This form k satisfies (6.1.9). For, if $\chi = \phi \otimes \psi$, where $\phi \in C_c^\infty(K_a)$ and $\psi \in C_c^\infty(K_a')$, then (6.1.16) gives, by the definition (6.1.9) of B,

$$\langle k, \phi\otimes\psi\rangle = \sum_{Z^n \times Z^m} \langle \mu\psi_h, \phi_g\rangle, \tag{6.1.17}$$

where

$$\phi_g = \hat{\phi}_g \rho(x)E(g\cdot x), \quad \psi_h = \hat{\psi}_h \sigma(y)E(h\cdot y)$$

and

$$\hat{\phi}_g = b^{-n}\int_{K_b} \phi(x)E(-g\cdot x)\,\mathrm{d}x, \quad \hat{\psi}_h = b^{-m}\int_{K_b'} \psi(y)E(-h\cdot y)\,\mathrm{d}y$$

are the Fourier coefficients of ϕ and ψ, respectively. These Fourier coefficients are rapidly decreasing, as one can obtain estimates similar to (6.1.14) by partial integration. So the two Fourier series converge in $C^\infty(\mathbf{R}^n)$ and in $C^\infty(\mathbf{R}^m)$, respectively, and the convergence is absolute. One can therefore sum first over g, since each $\mu\psi_h$ is a distribution, and then over h, since μ is sequentially continuous, to conclude that (6.1.17) reduces to (6.1.9).

The construction can evidently be carried out for any $a > 0$. One can therefore apply it to obtain a sequence k_j of continuous linear forms on $K_j \times K_j'$, where $j = 1, 2, \ldots$. As these all satisfy (6.1.9), Theorem 4.3.1 implies that the restriction of k_l to $K_j \times K_j'$ is equal to k_j if $l \geqslant j$. Hence there is a unique

$k \in \mathscr{D}'(\mathbf{R}^n \times \mathbf{R}^m)$ such that $k = k_j$ on $K_j \times K_j', j = 1, 2, \ldots$, which satisfies (6.1.9). The theorem is proved.

6.2. Regular kernels

There is a class of distribution kernels which is important in applications. Before discussing it, we need a definition.

Definition 6.2.1. Let $X \subset \mathbf{R}^n$ and $Y \subset \mathbf{R}^m$ be open sets. A map k: $C_c^\infty(Y) \to C^\infty(X)$ is called continuous if, for any two compact sets $K \subset X$ and $K' \subset Y$, and any multi-index α, there is a $C \geqslant 0$ and an integer $N \geqslant 0$ such that

$$\sup \{|\partial^\alpha kf(x)| : x \in K\} \leqslant C \sum_{|\beta| \leqslant N} \sup |\partial^\beta f|, \quad f \in C_c^\infty(K'). \tag{6.2.1}$$

Note. An equivalent condition is that k maps sequences converging to 0 in $C_c^\infty(Y)$ to sequences converging to 0 in $C^\infty(X)$; the proof is left as an exercise. Definition 6.2.1 should be compared with Definition 2.8.1.

Theorem 6.2.1. Suppose that the map ${}^tk$ generated by a Schwartz kernel $k \in \mathscr{D}'(X \times Y)$ is a continuous map $C_c^\infty(X) \to C^\infty(Y)$. Then the map k extends to a map $\mathscr{E}'(Y) \to \mathscr{D}'(X)$ which is sequentially continuous in the following sense: if a sequence $(u_j)_{1 \leqslant j < \infty}$ converges to $u \in \mathscr{E}'(Y)$ in $\mathscr{D}'(Y)$, and the supports of the u_j are in a fixed compact set, then $ku_j \to ku$ in $\mathscr{D}'(X)$ as $j \to \infty$.

Proof. This is similar to the proof of Theorem 2.8.1. Let $u \in \mathscr{E}'(Y)$; define $\tilde{k}u$ by

$$\langle \tilde{k}u, \phi \rangle = \langle u, {}^tk\phi \rangle, \quad \phi \in C_c^\infty(X). \tag{6.2.2}$$

As the composite of the continuous maps ${}^tk$: $C_c^\infty(X) \to C^\infty(Y)$ and u: $C^\infty(Y) \to \mathbf{C}$, $\tilde{k}u$ is a distribution. The sequential continuity of $\tilde{k}$: $\mathscr{E}'(Y) \to \mathscr{D}'(X)$ is also immediate from (6.2.2).

To show that $\tilde{k}$ extends k, take $u = \psi \in C_c^\infty(Y)$. Then (6.2.2) becomes

$$\langle \tilde{k}\psi, \phi \rangle = \langle \psi, {}^tk\phi \rangle = \int \psi(y)\, {}^tk\phi(y)\, dy.$$

Hence

$$\langle \tilde{k}\psi, \phi \rangle = \langle {}^tk\phi, \psi \rangle = \langle k\psi, \phi \rangle, \quad \phi \in C_c^\infty(X)$$

by (6.1.7), and we are done.

We can now obviously drop the tilde, and write k for the map defined by (6.2.2). The case most frequently encountered in applications is that of a kernel for which both k and ${}^tk$ are continuous maps $C_c^\infty(Y) \to C^\infty(X)$ and $C_c^\infty(X) \to C^\infty(Y)$, respectively. Such a kernel will be called a *regular kernel*. Theorem 6.2.1 then has the following corollary:

Corollary 6.2.1. If $k \in \mathscr{D}'(X \times Y)$ is a regular kernel, then the maps k and ${}^tk$ extend to sequentially continuous maps $\mathscr{E}'(Y) \to \mathscr{D}'(X)$ and $\mathscr{E}'(X) \to \mathscr{D}'(Y)$, respectively.

Let us now consider maps. A continuous map $\mu\colon C_c^\infty(Y) \to C^\infty(X)$ is evidently also a continuous map $C_c^\infty(Y) \to \mathscr{D}'(X)$. By Theorem 6.1.1 it is therefore generated by a Schwartz kernel; let us call this k_μ. The next theorem gives a rule for computing the kernel of μ.

Theorem 6.2.2. The Schwartz kernel of a continuous map $\mu\colon C_c^\infty(Y) \to C^\infty(X)$ is

$$\langle k_\mu, \chi \rangle = \int \mu\chi(x, \cdot)(x)\, dx, \quad \chi \in C_c^\infty(X \times Y), \tag{6.2.3}$$

where $\mu\chi(x, \cdot)(x)$ is the value at $x \in X$ of the image of $y \mapsto \chi(x, y)$ under μ.

Proof. It is sufficient to prove this when $\chi \in C_c^\infty(K_a \times K_a')$, where the notation is as in the proof of Theorem 6.1.1; the general case can be reduced to this by a partition of unity. The kernel k_μ is then given by (6.1.16), which becomes, in the present case,

$$\langle k_\mu, \chi \rangle = \sum_{\mathbf{Z}^n \times \mathbf{Z}^m} \int \hat{\chi}_{g,h} \phi_g(x) \mu\psi_h(x)\, dx, \tag{6.2.4}$$

where

$$\phi_g = \rho(x) E(g \cdot x), \quad \psi_h = \sigma(y) E(h \cdot y).$$

As supp $\phi_g \subset$ supp ρ and supp $\psi_h \subset$ supp σ, both of which are fixed compact sets, it follows from (6.2.1) that

$$|\phi_g(x)\mu\psi_h(x)| \leqslant C \sum_{|\alpha| \leqslant N} \sup |\partial^\alpha(\rho(y)E(h \cdot y))|$$

for some $C, N \geqslant 0$. Hence there is a constant C_1 such that

$$|\phi_g(x)\mu\psi_h(x)| \leqslant C_2(1 + |h|)^N.$$

It thus follows from (6.1.14), with M taken sufficiently large, that the series

$$\sum_{\mathbf{Z}^n \times \mathbf{Z}^m} \hat{\chi}_{g,h} \phi_g(x) \mu\psi_h(x)$$

converges uniformly to $\mu\chi(x, \cdot)(x)$. So one can reverse the order of summation and integration in (6.2.4), and this gives (6.2.3).

We now consider two examples. First, let us take $Y = X$, and let μ be a differential operator with C^∞ coefficients,

$$P(x, \partial) = \sum_{|\alpha| \leqslant m} a_\alpha(x) \partial^\alpha, \tag{6.2.5}$$

defined on X. (This is actually a map $C_c^\infty \to C_c^\infty$.) Then (6.2.3) becomes

$$\langle k_P, \chi\rangle = \int P(y, \partial_y)\chi(x,y)|_{y=x}\, dx, \quad \chi \in C_c^\infty(X \times X). \tag{6.2.6}$$

One can express this in terms of the kernel $\delta(x-y) \in \mathscr{D}'(X \times X)$ of the identity map, which is

$$\langle \delta(x-y), \chi(x,y)\rangle = \int \chi(x,x)\, dx, \quad \chi \in C_c^\infty(X \times X). \tag{6.2.7}$$

For (6.2.6) is clearly equivalent to

$$k_P = {}^tP(y, \partial_y)\delta(x-y), \tag{6.2.8}$$

where ${}^tP$ is the adjoint of P, given by (2.6.3).

In particular, if $m=0$ and $P=g \in C^\infty(X)$, then $\psi \mapsto P\psi$ reduces to the multiplication map $\psi \mapsto g\psi$, and one has $k_g = g(x)\delta(x-y)$; and if $P=\partial_i, i=1,2,\ldots,n$, then the kernel is $-(\partial/\partial y_i)\delta(x-y)$.

As a second example, we shall use Theorem 6.2.2 and Corollary 6.2.1 to give an alternative treatment of the convolution of two distributions, one of which has compact support. Take $X=Y=\mathbf{R}^n$, and let μ be defined by

$$\mu\psi = u * \psi = \langle u(y), \psi(x-y)\rangle, \quad \psi \in C_c^\infty(\mathbf{R}^n), \tag{6.2.9}$$

where u is a given distribution. To prove that μ is a continuous map $C_c^\infty(\mathbf{R}^n) \to C^\infty(\mathbf{R}^n)$, one notes first that $\mu\psi \in C^\infty(\mathbf{R}^n)$ by Theorem 4.1.1, and that this theorem also gives

$$\partial^\alpha(\mu\psi(x)) = \langle u(y), \partial^\alpha\psi(x-y)\rangle.$$

Now, let K and K' be compact sets in $\mathbf{R}^n$, and suppose that $x \in K$, supp $\psi \in K'$. Then supp $(y \mapsto \psi(x-y)) \subset K - K'$ which is bounded and hence relatively compact. (It is actually a compact set, by a result established in the proof of Theorem 5.1.2, but we do not need this.) Hence there is a semi-norm estimate for u which gives

$$\sup\{|\partial^\alpha(\mu\psi(x))| : x \in K\} \leqslant C \sum_{|\alpha| \leqslant N} \sup |\partial^\alpha\psi|, \quad \psi \in C_c^\infty(K'),$$

and this implies (6.2.1).

Let k_u be the kernel of the map (6.2.9); by Theorem 6.2.2, one has

$$\langle k_u, \chi\rangle = \int \langle u(y), \chi(x, x-y)\rangle\, dx, \quad \chi \in C_c^\infty(\mathbf{R}^n \times \mathbf{R}^n). \tag{6.2.10}$$

(Essentially, this 'is' the distribution $u(x-y) \in \mathscr{D}'(X \times X)$.) So

$$\langle {}^tk_u\phi, \psi\rangle = \langle k_u, \phi \otimes \psi\rangle = \int \langle u(y), \phi(x)\psi(x-y)\rangle\, dx$$

$$= \int \langle u(y), \phi(x+y)\rangle \psi(x)\, dx,$$

whence

$${}^t k_u \phi = \langle u(y), \phi(x+y) \rangle = \check{u} * \phi, \quad \phi \in C_c^\infty(\mathbf{R}^n), \tag{6.2.11}$$

where $\check{u}$ is u reflected in the origin. Thus ${}^t k_u$ is also a continuous map $C_c^\infty(\mathbf{R}^n) \to C^\infty(\mathbf{R}^n)$, k_u is a regular kernel, and (6.2.9) extends to a sequentially continuous map $\mathscr{E}'(\mathbf{R}^n) \to \mathscr{D}'(\mathbf{R}^n)$. Of course, this is just the convolution map $v \mapsto u * v$.

By (5.1.6), the convolution map commutes with translations. Our final theorem in this subsection says that convolution is characterized by this property.

Theorem 6.2.3. A continuous map μ: $C_c^\infty(\mathbf{R}^n) \to C^\infty(\mathbf{R}^n)$ commutes with translation if and only if $\mu\psi = u * \psi$ for some $u \in \mathscr{D}'(\mathbf{R}^n)$.

Proof. We have already proved the necessity of the condition. Conversely, we observe that (6.2.1) with $\alpha = 0$ and $K = \{0\}$ shows that $\psi \mapsto \mu\psi(0)$ is a distribution; let us call this v. As μ commutes with (all) translations, one now has, for any $h \in \mathbf{R}^n$,

$$\mu\psi(h) = \tau_{-h} \circ \mu\psi(0) = \mu \circ \tau_{-h}\psi(0) = \langle v, \tau_{-h}\psi \rangle.$$

So $\mu\psi = u * \psi$ where $u = \check{v}$, and we are done.

Remark. In the proof of sufficiency, we have only used (6.2.1) when $\alpha = 0$ and K is a point; the theorem implies that this in turn ensures the validity of (6.2.1) quite generally.

6.3. Fundamental kernels of differential operators

It has already been pointed out that fundamental solutions of differential operators with constant coefficients play an important part in the theory of such operators. In general, a similar part is played by fundamental kernels.

Let $P(x, \partial)$ be a linear differential operator with C^∞ coefficients, defined on an open set $X \subset \mathbf{R}^n$. A distribution $E \in \mathscr{D}'(X \times X)$ is called a *right fundamental kernel* of P if the map E which it generates as a Schwartz kernel is a right inverse of P,

$$PE\psi = \psi, \quad \psi \in C_c^\infty(X). \tag{6.3.1}$$

Likewise, $E' \in \mathscr{D}'(X \times X)$ is called a *left fundamental kernel* of P if the corresponding map E' is a left inverse of P,

$$E'P\psi = \psi, \quad \psi \in C_c^\infty(X). \tag{6.3.2}$$

Right fundamental kernels give existence theorems, left fundamental kernels give uniqueness theorems; these, however, usually need an extension of the domain of E' to be useful.

Let us consider (6.3.1) in more detail: it means that

$$\langle E, {}^tP\phi \otimes \psi \rangle = \langle \delta(x-y), \phi \otimes \psi \rangle, \quad \phi, \psi \in C_c^\infty(X), \tag{6.3.3}$$

where ${}^tP$ is the adjoint of P, and $\delta(x-y)$ is the kernel of the identity map on

$C_c^\infty(X)$, (6.2.7). This identity extends by continuity to

$$\langle E(x,y), {}^tP(x,\partial_x)\chi(x,y)\rangle = \langle \delta(x-y), \chi(x,y)\rangle, \quad \chi \in C_c^\infty(X\times X)$$

which is nothing other than

$$P(x,\partial_x)E(x,y) = \delta(x-y). \tag{6.3.4}$$

As this equation implies (6.3.1), it is an alternative characterization of right fundamental kernels of P.

It also follows from (6.3.3) that

$$ {}^tE\,{}^tP\phi = \phi, \quad \phi \in C_c^\infty(X), \tag{6.3.5}$$

So the transpose of a right fundamental kernel of P is a left fundamental kernel of ${}^tP$.

The equation (6.3.2) can be discussed in the same manner. One finds that it is equivalent to

$$ {}^tP(y,\partial_y)E'(x,y) = \delta(x-y), \tag{6.3.6}$$

and that ${}^tE'$ is a right fundamental kernel of ${}^tP$.

It can of course happen that a kernel is both left and right fundamental, so that both (6.3.1), (6.3.2) or (6.3.4), (6.3.6) hold for one and the same $E \in \mathscr{D}'(X\times X)$.

An important case, which occurs frequently, is that of a (right) fundamental kernel which is a regular kernel. Suppose, then, that E satisfies (6.3.1) and that both E and ${}^tE$ are continuous maps $C_c^\infty \to C^\infty$. Then one can obtain a solution of

$$Pu = f, \quad f \in \mathscr{E}'(X). \tag{6.3.7}$$

Indeed, following the proof of Theorem 6.2.1, one only has to set

$$\langle u, \phi\rangle = \langle f, {}^tE\phi\rangle, \quad \phi \in C_c^\infty(X). \tag{6.3.8}$$

For this defines a distribution, and if $\phi \in C_c^\infty(X)$ then

$$\langle Pu, \phi\rangle = \langle u, {}^tP\phi\rangle = \langle f, {}^tE\,{}^tP\phi\rangle = \langle f, \phi\rangle,$$

by (6.3.5). One can obtain the same u as a limit in $\mathscr{D}'(X)$,

$$u = \lim_{\epsilon\to 0+} Ef_\epsilon$$

where the f_ϵ are regularizations of f, converging to f in $\mathscr{D}'(X)$ as $\epsilon \to 0+$ and supported in a compact neighbourhood of the support of f.

One can also define fundamental solutions of a differential operator P. Let y be a point of X, and let δ_y be the y-translate of the Dirac distribution,

$$\langle \delta_y, \phi\rangle = \phi(y), \quad \phi \in C_c^\infty(X). \tag{6.3.9}$$

A distribution $E_y \in \mathscr{D}'(X)$, depending on y as a parameter, is called a *fundamental solution* of P if

$$P(x,\partial_x)E_y = \delta_y, \tag{6.3.10}$$

which means that

$$\langle E_y, {}^tP\phi\rangle = \phi(y), \quad \phi \in C_c^\infty(X). \tag{6.3.11}$$

Suppose that $y \mapsto \langle E_y, \phi\rangle$ is in $C^\infty(X)$, and that the map $\phi \mapsto \langle E_y, \phi\rangle$ is a continuous map $C_c^\infty(X) \to C^\infty(X)$. Then the kernel $E(x,y)$ of this map is a right fundamental kernel of P. For, by Theorem 6.2.2 one has then

$$\langle E, \chi\rangle = \int \langle E_y(x), \chi(x,y)\rangle \, dy, \quad \chi \in C_c^\infty(X \times X) \tag{6.3.12}$$

and it follows from (6.3.11) that this gives (6.3.1). In particular, if P has constant coefficients and is defined on $\mathbf{R}^n$, one can take $y = 0$ and derive a kernel from $E_y = \tau_y E_0$. It is an easy exercise to deduce from (6.3.12) and (6.2.10) that (6.3.8) then becomes $u = E_0 * f$, in accordance with the general scheme (5.3.11)-(5.3.13) for convolution equations.

Exercises

6.1. Show that, if $k \in C^\infty(\mathbf{R}^n \times \mathbf{R}^n)$ acts as a Schwartz kernel, then the map $\psi \to k\psi$ extends to a map $\mathscr{E}'(\mathbf{R}^n) \to C^\infty(\mathbf{R}^n)$.

6.2. Let $X \subset \mathbf{R}^n$ and $Y \subset \mathbf{R}^m$ be open sets, and let $k \in \mathscr{D}'(X \times Y)$ be a distribution kernel. Show that, if $\psi \in C_c^\infty(Y)$, then

$\operatorname{supp} k \circ \operatorname{supp} \psi = \{x\text{: there is a } y \in \operatorname{supp} \psi \text{ such that } (x,y) \in \operatorname{supp} k\}$

is a closed set, and that $\operatorname{supp} k\psi \subset \operatorname{supp} k \circ \operatorname{supp} \psi$.

Deduce that, if the projections $X \times Y \to X$ and $X \times Y \to Y$ are proper maps when restricted to the support of k, then $\psi \to k\psi$ is a map $C_c^\infty(Y) \to \mathscr{E}'(X)$ and extends to a map $C^\infty(Y) \to \mathscr{D}'(X)$.

6.3. (Peetre's theorem.) Let $k \in \mathscr{D}'(\mathbf{R}^n \times \mathbf{R}^n)$, and suppose that the map k which it generates as a Schwartz kernel preserves support: $\operatorname{supp} k\psi \subset \operatorname{supp} \psi$ for all $\psi \in C_c^\infty(\mathbf{R}^n)$. Prove that locally (that is, in any bounded subset of $\mathbf{R}^n \times \mathbf{R}^n$), the map k is a differential operator.

(Suggestion: look first at Exercise 4.6.)

6.4. If $\mu\colon C_c^\infty(\mathbf{R}^n) \to C^\infty(\mathbf{R}^n)$ is a continuous map, is it necessary to appeal to the theorem that a separately continuous bilinear form on the product of two Fréchet spaces is jointly continuous, in order to prove the kernel theorem for μ, and to establish (6.2.3)?

6.5. *Define* the convolution of $u \in \mathscr{D}'(\mathbf{R}^n)$ and $\psi \in C_c^\infty(\mathbf{R}^n)$ as $u * \psi = \langle u(y), \psi(x-y)\rangle$; *prove* that, if $v \in \mathscr{E}'(\mathbf{R}^n)$, then there is a $w \in \mathscr{D}'(\mathbf{R}^n)$ such that

$$v * (u * \psi) = w * \psi, \quad \psi \in C_c^\infty(\mathbf{R}^n).$$

(This is another way of defining the convolution of two distributions, one of which has compact support.)

6.6. Let

$$P(x, \partial) = a_0(x)\partial^m + a_1(x)\partial^{m-1} + \ldots + a_m(x)$$

be a linear differential operator with C^∞ coefficients, defined on an open interval $J \subset \mathbf{R}$, and assume that $a_0 \neq 0$ on J. Let y be a point of J, and let $U(x,y)$ be the solution of the boundary value problem

$$P(x, \partial_x)U = 0 \quad \text{on } J, \quad \partial_x^k U|_{x=y} = 0, \quad 0 \leqslant k < m-1, \quad \partial_x^{m-1} U|_{x=y} = 1.$$

Put

$$E(x, y) = U(x, y)H(x - y),$$

where H is the Heaviside function. Prove that:

(i) E is a right fundamental kernel of P;

(ii) the map $\psi \mapsto E\psi$ is well defined if the support of ψ is bounded on the left, that is to say supp $\psi \subset J \cap [c, \infty)$ for some $c \in J$, and that then

$$\text{supp } E\psi \subset \{x \in J\colon \text{there is a } y \in \text{supp } \psi \text{ such that } x \geqslant y\};$$

(iii) E is also a left fundamental kernel of P;

(iv) if $f \in \mathscr{E}'(J)$ is given, then there is one and only one $u \in \mathscr{D}'(J)$ such that

$$Pu = f, \quad \text{supp } u \text{ is bounded on the left.}$$

(Suggestion. For (iii), appeal to the uniqueness theorem for the boundary value problem

$$P\psi = \phi, \quad \phi \in C_c^\infty(J), \quad \text{supp } \psi \text{ is bounded on the left.}$$

See [1, Ch. X] for existence, uniqueness and regularity theorems for ordinary differential equations.)

7 COORDINATE TRANSFORMATIONS AND PULLBACKS

If $f: X \to Y$ is a map on an open set $X \subset \mathbf{R}^n$ whose range is contained in an open set $Y \subset \mathbf{R}^m$, then any function on Y can be pulled back to a function on X by composition with f. If $f \in C^\infty(X \to Y)$ (this means that f is a C^∞ map $X \to Y$), one obtains in this way a vector space homomorphism $f^*: C^\infty(Y) \to C^\infty(X)$, given by $f^*u = u \circ f$.

To extend this operation to distributions, one can use the fact, implied by Theorem 5.2.3, that $C^\infty(Y)$ is dense in $\mathscr{D}'(Y)$. The pullback map $f^*: \mathscr{D}'(Y) \to \mathscr{D}'(X)$ then exists if composition with f maps sequences converging in $\mathscr{D}'(Y)$ to sequences converging in $\mathscr{D}'(X)$. If one also requires f^* to be a sequentially continuous map $\mathscr{D}'(Y) \to \mathscr{D}'(X)$, then this condition is both necessary and sufficient for the existence of a unique pullback map.

We shall now consider some cases of maps $f \in C^\infty(X \to Y)$ where f^* can be defined on all of $\mathscr{D}'(Y)$, and an example where it cannot be defined for all distributions on Y, but can be constructed for some distributions that are of interest in applications.

7.1. Diffeomorphisms

A map $f \in C^\infty(X \to Y)$ is called a diffeomorphism if it is a bijection, and the inverse map $f^{-1}: Y \to X$ is also C^∞. For example, if $X = Y = \mathbf{R}$, then $x \mapsto y = cx$, where $0 \neq c \in \mathbf{R}$, is a diffeomorphism, but $x \mapsto y = x^3$ is a bijection which is not a diffeomorphism. For a diffeomorphism, one necessarily has $m = n$. Furthermore, the Jacobian, $\det f'(x)$, does not vanish on X; here the derivative f' of f is the $n \times n$ matrix with the entries $\partial f_i/\partial x_j$, which is invertible when f is a diffeomorphism. The pullback of a distribution by a diffeomorphism exists, and is easy to compute:

Theorem 7.1.1. Let X and Y be open sets in $\mathbf{R}^n$, and let $f: X \to Y$ be a diffeomorphism. Then the pullback $f^*u \in \mathscr{D}'(X)$ of any $u \in \mathscr{D}'(Y)$ exists. It is given by

$$\langle f^*u, \phi \rangle = \langle u(y), g^*\phi(y) |\det g'(y)| \rangle, \quad \phi \in C_c^\infty(X), \tag{7.1.1}$$

where $g = f^{-1}$, and is a sequentially continuous map $\mathscr{D}'(Y) \to \mathscr{D}'(X)$.

Proof. As g is continuous, it maps compact sets in Y to compact sets in X. Also, one has $\det g' = g^*(1/\det f') \neq 0$ on Y. From these two facts, and the chain rule, it follows that $\phi \to g^*\phi|\det g'|$ is a continuous map $C_c^\infty(X) \to C_c^\infty(Y)$, according to Definition 2.8.1. By Theorem 2.8.1, the equation (7.1.1) therefore defines a sequentially continuous map $f^*: \mathscr{D}'(Y) \to \mathscr{D}'(X)$. Furthermore, if $u \in C^\infty(Y)$, then (7.1.1) becomes

$$\langle f^*u, \phi\rangle = \int u(y) g^*\phi(y) |\det g'(y)|\, dy$$

$$= \int u \circ f(x)\phi(x)\, dx, \quad \phi \in C_c^\infty(X),$$

by the rule for changing variables in an integral. So $f^*u = u \circ f$ in this case, and we are done.

Corollary 7.1.1. The derivatives of f^*u can be computed by the chain rule.

Proof. This is elementary when $u \in C^\infty(Y)$, and extends to distributions by sequential continuity.

One can obviously think of a diffeomorphism as a coordinate transformation, with $\mathbf{R}^n$ considered as a C^∞ manifold. Indeed, Theorem 7.1.1 is the first step in the extension of distribution theory to manifolds, but this is a matter which will not be taken up in this book.

We have already met two examples: translations, and the maps $x \mapsto Ax$ where A is a non-singular $n \times n$ matrix. This is clear when one compares (7.1.1) with (4.2.3) and (4.2.7). Another example is the pullback of the Dirac distribution under a diffeomorphism. Write, for any $h \in Y$,

$$\langle \delta_h, \phi\rangle = \phi(h), \quad \phi \in C_c^\infty(Y). \tag{7.1.2}$$

Then a simple computation shows that (7.1.1) gives

$$f^*\delta_h = |\det f'(g)|^{-1}\delta_g, \quad \text{where } g = f^{-1}(h). \tag{7.1.3}$$

7.2. The pullback of a distribution by a function

When $m = 1$, one may simply take $Y = \mathbf{R}$, and f is just a real valued C^∞ function on X. Let us assume that

$$f' = (\partial_1 f, \ldots, \partial_n f) \neq 0 \quad \text{on } X. \tag{7.2.1}$$

Then the pullback map f^* can be defined.

First, take $x = \mathbf{R}^n$ so that $f(x) = y$. Then $(f^*u)(x) = (u \circ f)(x)$ when $u \in C^\infty(\mathbf{R})$. So, for any $\phi \in C_c^\infty(X)$,

$$\langle f^*u, \phi\rangle = \int u(x_n)\phi(x', x_n)\, dx = \left\langle u(t), \int \phi(x', t)\, dx' \right\rangle \tag{7.2.2}$$

where $x' = (x_1, \ldots, x_{n-1})$. This amounts to

$$f^*u = 1(x') \otimes u(x_n) \tag{7.2.3}$$

where $1(x')$ is the constant function, equal to unity, on $\mathbf{R}^{n-1}$. Clearly, (7.2.3) defines the pullback map for all $u \in \mathscr{D}'(\mathbf{R})$.

The general case can be reduced to this example by a coordinate transformation. To write the result in a coordinate independent form, one can use the following device. For any $\phi \in C_c^\infty(X)$ and any $t \in \mathbf{R}$, set

$$\phi_f(t) = \frac{\partial}{\partial t} \int_{f(x)<t} \phi(x)\, \mathrm{d}x. \tag{7.2.4}$$

It will presently be shown that $\phi_f(t) \in C_c^\infty(\mathbf{R})$. Assuming this, let us compute $\langle u, \phi_f \rangle$ when $u \in C_c^\infty(\mathbf{R})$. An integration by parts gives

$$\int u(t)\phi_f(t)\, \mathrm{d}t = -\int \partial u(t)\, \mathrm{d}t \int_{f(x)<t} \phi(x)\, \mathrm{d}x$$

$$= -\int \phi(x)\, \mathrm{d}x \int_{f(x)}^{\infty} \partial u(t)\, \mathrm{d}t = \int u \circ f(x)\phi(x)\, \mathrm{d}x,$$

so that, writing f^*u for $u \circ f$, one has

$$\langle f^*u, \phi \rangle = \langle u(t), \phi_f(t) \rangle, \quad \phi \in C_c^\infty(X). \tag{7.2.5}$$

We now have:

***Theorem** 7.2.1.* Let $X \subset \mathbf{R}^n$ be an open set, let $f \in C^\infty(X)$ be real valued, and assume that $f'(x) \neq 0$ on X. Then the pullback $f^*u \in \mathscr{D}'(X)$ of any $u \in \mathscr{D}'(\mathbf{R})$ exists; it is given by (7.2.4) and (7.2.5), and $f^*: \mathscr{D}'(\mathbf{R}) \to \mathscr{D}'(X)$ is sequentially continuous.

Proof. By hypothesis, (7.2.1) holds, and so at least one of the $\partial_i f$ is $\neq 0$ at each point $y \in X$. By the implicit function theorem [1, Ch X], there is thus an open neighbourhood U_y of y in X, and a diffeomorphism $x \mapsto \xi = h_y(x)$ of U_y on an open set $h_y U_y \subset \mathbf{R}^n$, such that $\xi_n = f(x)$. For example, if $\partial_n f(y) \neq 0$, one can determine such a h_y by inverting

$$\xi_1 = x_1, \ldots, \xi_{n-1} = x_{n-1}, \quad \xi_n = f(x)$$

locally. If now $\phi \in C_c^\infty(U_y)$, then (7.2.4) becomes

$$\phi_f(t) = \frac{\partial}{\partial t} \int_{\xi_n < t} g_y^*\phi(\xi) |\det g_y'(\xi)|\, \mathrm{d}\xi$$

where $g_y = h_y^{-1}$. Hence

$$\phi_f(t) = \int g_y^*\phi(\xi', t) |\det g_y'(\xi', t)|\, \mathrm{d}\xi', \quad \phi \in C_c^\infty(U_y), \tag{7.2.6}$$

with $\xi' = (\xi_1, \ldots, \xi_{n-1})$. So (7.2.5), restricted to U_y, is equivalent to

$$f^*u = |\det g_y'(\xi)|\, 1(\xi') \otimes u(\xi_n) \tag{7.2.7}$$

which is similar to (7.2.3).

We have thus constructed a pullback map $f_y^*: \mathscr{D}'(\mathbf{R}) \to \mathscr{D}'(U_y)$ which is equivalent to (7.2.5). This shows that the compatibility conditions of Theorem 1.4.3 hold, so that the f_y^*u piece together to give a distribution f^*u for any $u \in \mathscr{D}'(\mathbf{R})$. As it is immediate from (7.2.7) and a partition of unity argument that $\phi_f \in C_c^\infty(\mathbf{R})$, it is clear that f^*u is given by (7.2.5). The sequential continuity of f^* is now obvious, and so we are done.

Corollary 7.2.1. The derivatives of f^*u can be computed by the chain rule.

Proof. This is obvious when $u \in C^\infty(\mathbf{R})$, and extends to distributions by the sequential continuity of f^*.

Let $s \in \mathbf{R}$, and let us write $\delta_s(t)$ for the translate $\tau_s\delta$. Then (7.2.5) gives

$$\langle f^*\delta_s, \phi\rangle = \phi_f(s), \quad \phi \in C_c^\infty(X), \tag{7.2.8}$$

which is virtually another definition of ϕ_f. We can derive yet another form by an appeal to the divergence theorem. Let H be the Heaviside function, and write $H(f(x)-s)$ for $f^*H(t-s)$. Then Corollary 7.2.1 gives, for $i=1, \ldots, n$,

$$\langle(\partial_i f) f^*\delta_s, \phi\rangle = \langle \partial_i H(f(x)-s), \phi\rangle, \quad \phi \in C_c^\infty(X). \tag{7.2.9}$$

On the other hand, if $\psi \in C^\infty(\mathbf{R})$ is such that $0 \leqslant \psi \leqslant 1$, $\psi = 0$ for $t < -1$, and $\psi = 1$ for $t > 1$, then $\psi(t/\epsilon) \to H(t)$ boundedly as $\epsilon \to 0+$. Hence

$$\langle H(f(x)-s), \phi(x)\rangle = \int_{f(x)>s} \phi(x)\, \mathrm{d}x.$$

It follows that

$$\langle \partial_i H(f(x)-s), \phi\rangle = -\int_{f(x)>s} \partial_i \phi\, \mathrm{d}x = \int_{\Sigma_s} \phi N_i\, \mathrm{d}\Sigma_s,$$

by the divergence theorem; here Σ_s is the surface $\{x : f(x) = s\}$, $\mathrm{d}\Sigma_s$ is the usual Euclidean surface measure on Σ_s, and $N = f'/|f'|$ is the unit normal to Σ_s. in the direction of increasing f. If we substitute this in (7.2.9), replace ϕ by $\phi\partial_i f/|f'|^2$, and sum over $i = 1, \ldots, n$, we obtain by comparison with (7.2.8) that

$$\phi_f(s) = \int_{\Sigma_s} \phi(x)/|f'(x)|\, \mathrm{d}\Sigma_s. \tag{7.2.10}$$

This form of ϕ_f brings out the reason for making the hypothesis (7.2.1) very clearly.

It has been assumed for simplicity that f is a map $X \to \mathbf{R}$. The reader should have no difficulty in verifying that Theorem 7.2.1 and its consequences hold equally for any $f \in C_c^\infty(X \to Y)$ when Y is an open subset of $\mathbf{R}$, and (7.2.1) is assumed.

Let us return briefly to the general case. A map $f \in C^\infty(X \to Y)$ is called a *submersion* if $m \leqslant n$, and $f'(x)$ is surjective for all $x \in X$. (This just means that the matrix $(\partial f_i / \partial x_j)$ has maximal rank m for all $x \in X$.)

Theorem 7.2.2. Let $X \subset \mathbf{R}^n$ and $Y \subset \mathbf{R}^m$ be open sets, and let $f\colon X \to Y$ be a submersion. Then there is a unique, sequentially continuous pullback map $f^*\colon \mathcal{D}'(Y) \to \mathcal{D}'(X)$.

Proof. As this is a simple modification of the proof of Theorem 7.2.1, it will only be sketched; the details are left to the reader.

If $m = n$, then a submersion is locally a diffeomorphism, by the inverse function theorem. So an appeal to the Heine-Borel theorem, and a partition of unity argument, reduce the proof of Theorem 7.2.2 to that of Theorem 7.1.1.

If $m < n$, let us write

$$\{f(x) < t\} = \{x\colon f_j(x) < t_j, \quad j = 1, \ldots, m\}$$

and set, for any $\phi \in C_c^\infty(X)$ and $t \in Y$

$$\phi_f(t) = \frac{\partial^m}{\partial t_1 \ldots \partial t_m} \int_{f(x) < t} \phi(x)\, \mathrm{d}x. \tag{7.2.11}$$

Let $\bar{x}$ be a point of X. Since f is a submersion, it follows from the implicit function theorem that there is a neighbourhood $\bar{U} \subset X$ of $\bar{x}$ and a diffeomorphism $\bar{h}\colon \bar{U} \to hU$ such that, if we write $\xi = \bar{h}(x)$ as (ξ', ξ'') where $\xi' \in \mathbf{R}^{n-m}$ and $\xi'' \in \mathbf{R}^m$, then $\xi'' = f(x)$. Hence, with $\bar{g} = \bar{h}^{-1}$, one has

$$\phi_f(\xi'') = \int \bar{g}^* \phi(\xi', \xi'') |\det \bar{g}'(\xi', \xi'')|\, \mathrm{d}\xi',$$

and from this it is easy to argue that

$$\bar{f}^* u = |\det \bar{g}'(\xi)| 1(\xi') \otimes u(\xi'')$$

gives the restriction of the pullback map $\bar{f}^*\colon \mathcal{D}'(Y) \to \mathcal{D}'(\bar{U})$. Theorem 1.4.3 again implies that these distributions $\bar{f}^* u$ piece together to give $f^* u \in \mathcal{D}'(X)$, and it is also immediate that

$$\langle f^* u, \phi \rangle = \langle u, \phi_f \rangle, \quad \phi \in C_c^\infty(X). \tag{7.2.12}$$

This, in turn, shows that $f^*\colon \mathcal{D}'(Y) \to \mathcal{D}'(X)$ is a sequentially continuous map, and so the proof is complete.

One can give another representation of ϕ_f by means of exterior forms. Write

$$\omega = dx_1 \wedge \ldots \wedge dx_n. \tag{7.2.13}$$

Then one can determine an m-form ω_f such that

$$\omega = df_1 \wedge \ldots \wedge df_m \wedge \omega_f \tag{7.2.14}$$

The restriction of ω_f to a manifold $f(x)=t$ is unique, and one has

$$\phi_f(t) = \int_{f(x)=t} \phi\omega_f. \tag{7.2.15}$$

7.3. The wave equation on $\mathbf{R}^4$

The restriction to submersions in Theorems 7.2.1 and 7.2.2 can sometimes be relaxed to good effect for particular distributions. As an example of this, we shall now construct fundamental solutions of the wave equation on $\mathbf{R}^4$.

Write $x \in \mathbf{R}^4$ as (x', x_4), where $x' \in \mathbf{R}^3$, and take

$$f(x) = x_4^2 - |x'|^2. \tag{7.3.1}$$

The only critical point of f is the origin, so $f: \mathbf{R}^4 \setminus \{0\} \to \mathbf{R}$ is a submersion. Let $\phi \in C_c^\infty(\mathbf{R}^4 \setminus \{0\})$, and let us compute $f^*\delta = \phi_f(0)$. By (7.2.4), this is

$$\frac{\partial}{\partial t} \int_{-(|x'|^2+t)_+^{1/2}}^{(|x'|^2+t)_+^{1/2}} dx_4 \int \phi(x', x^4)\, dx' |_{t=0}$$

$$= \tfrac{1}{2} \int \phi(x', |x'|)|x'|^{-1}\, dx' + \tfrac{1}{2} \int \phi(x', -|x'|)|x'|^{-1}\, dx'. \tag{7.3.2}$$

Obviously, one can write this as

$$f^*\delta = \delta_+(f) + \delta_-(f), \tag{7.3.3}$$

where

$$\langle \delta_+(f), \phi \rangle = \tfrac{1}{2} \int \phi(x', |x'|)|x'|^{-1}\, dx' \tag{7.3.4}$$

and

$$\langle \delta_-(f), \phi \rangle = \tfrac{1}{2} \int \phi(x', -|x'|)|x'|^{-1}\, dx'. \tag{7.3.5}$$

As these two distributions are obtained from each other by reflection in $x_4 = 0$, it will be sufficient to consider (7.3.4). This is a measure carried on the *forward null cone*

$$\Gamma^+ = \{x : x_4 = |x'|\} \tag{7.3.6}$$

and obviously extends to a distribution on $\mathbf{R}^4$. We shall now show that it is a constant multiple of a fundamental solution of the d'Alembertian

$$P(\partial) = \partial_4^2 - \partial_1^2 - \partial_2^2 - \partial_3^2. \tag{7.3.7}$$

By Corollary 7.2.1, $Pf^*\delta$ can be computed on $\mathbf{R}^4\setminus\{0\}$ by the chain rule. A straightforward calculation, which is left to the reader, shows that $Pf^*\delta = 0$ on $\mathbf{R}^4\setminus\{0\}$. As the supports of $\delta_+(f)$ and $\delta_-(f)$ are disjoint then, it follows from (7.3.3) that one also has $P\delta_+(f) = 0$ on $\mathbf{R}^4\setminus\{0\}$. So the support of $P\delta_+(f)$ is the origin, and hence, for some $m<\infty$,

$$P\delta_+(f) = \sum_{|\alpha|\leq m} c_\alpha \partial^\alpha \delta \tag{7.3.8}$$

by Theorem 3.2.1. Now $\delta_+(f)$ is homogeneous of degree -2. For, if $t>0$, and ϕ is replaced by $t^{-4}\phi(x/t)$ in (7.3.4), then the second member becomes

$$\tfrac{1}{2}t^{-4}\int \phi(x'/t, |x'|/t)|x'|^{-1}\,dx'$$

$$= \tfrac{1}{2}t^2 \int \phi(x', |x'|)|x'|^{-1}\,dx' = t^2\langle \delta_+(f), \phi\rangle,$$

by the change of variable $x' \mapsto x't$. Thus $P\delta_+(f)$ is homogeneous of degree -4, and as $\partial^\alpha\delta$ is homogeneous of degree $(-4-|\alpha|)$, it follows that $c_\alpha = 0$ in (7.3.8) when $\alpha > 0$. Hence

$$P\delta_+(f) = c\delta, \quad c\in\mathbf{C}. \tag{7.3.9}$$

It only remains to determine the constant c. To this end, we observe that (7.3.4) extends to any test function whose support meets the forward null cone (7.3.6) in a compact set. So one can take $\phi = \rho(x_4)$, where $\rho\in C_c^\infty(\mathbf{R})$, and (7.3.4) then gives, in view of (7.3.9),

$$c\rho(0) = \langle \delta_+(f), P\rho\rangle = \tfrac{1}{2}\int |x'|^{-1}\partial^2\rho(|x'|)\,dx'.$$

Going over to polar coordinates this becomes

$$c\rho(0) = \tfrac{1}{2}\cdot 4\pi \int_0^\infty r\partial^2\rho(r)\,dr = 2\pi\rho(0),$$

whence $c = 2\pi$. Because of the symmetrical ('time reversed') behaviour of $\delta_-(f)$, we have therefore proved:

Theorem 7.3.1. The distributions E^+ and E^-, defined by

$$\langle E^+, \phi\rangle = \frac{1}{4\pi}\int \phi(x', |x'|)|x'|^{-1}\,dx', \quad \phi\in C_c^\infty(\mathbf{R}^4) \tag{7.3.10}$$

and

$$\langle E^-, \phi \rangle = \frac{1}{4\pi} \int \phi(x', -|x'|)|x'|^{-1}\, dx', \quad \phi \in C_c^\infty(\mathbf{R}^4) \tag{7.3.11}$$

are fundamental solutions of the d'Alembertian (7.3.7).

The support of E^+ is the forward null cone, and this property of E^+ has an important consequence:

Theorem 7.3.2. Let $v \in \mathscr{D}'(\mathbf{R}^4)$ be supported in $\{x_4 \geqslant 0\}$. Then the wave equation

$$Pu \equiv (\partial_4^2 - \partial_1^2 - \partial_2^2 - \partial_3^2)u = v \tag{7.3.12}$$

has a unique solution $u \in \mathscr{D}'(\mathbf{R}^4)$ with support in $x_4 \geqslant 0$. Also

$$\operatorname{supp} u \subset \operatorname{supp} v + \Gamma^+ = x\colon \text{there is a } y \in \operatorname{supp} v \text{ such that } x_4 = y_4 + |x' - y'|\}. \tag{7.3.13}$$

Proof. The solution in question is $u = E^+ * v$. This is a consequence of Theorem 5.3.1, and a special case of Exercise 5.5, so is left to the reader. The inclusion (7.3.13) follows from Theorem 5.3.2 (iii).

Taking $v = \delta$ in (7.3.12), one obtains

Corollary 7.3.1. E^+ is the unique fundamental solution of the d'Alembertian supported in $\{x_4 \geqslant 0\}$.

Theorem 7.3.2 evidently remains valid if $\operatorname{supp} v \subset \{x_4 \geqslant c\}$, for some $c \in \mathbf{R}$, with $\operatorname{supp} u \subset \{x_4 \geqslant c\}$. There are similar results for 'backward' solutions of the wave equation, constructed by means of E^-.

To compute $E^+ * v$, one has, with $\phi \in C_c^\infty(\mathbf{R}^4)$

$$\langle E^+ * v, \phi \rangle = \langle E^+(x) \otimes v(y), \phi(x+y) \rangle$$

$$= \frac{1}{4\pi} \left\langle v(y), \int \phi(x' + y', |x'| + y_4)|x'|^{-1}\, dx' \right\rangle$$

or

$$\langle E^+ * v, \phi \rangle = \frac{1}{4\pi} \left\langle v(y), \int \phi(x', y_4 + |x' - y'|)|x' - y'|^{-1}\, dx' \right\rangle. \tag{7.3.14}$$

If $v \in C^\infty(\mathbf{R}^4)$, then the second member becomes

$$\frac{1}{4\pi} \int \phi(x', y_4 + |x' - y'|) v(y) \frac{dx'\, dy'\, dy_4}{|x' - y'|}$$

$$= \frac{1}{4\pi} \int \phi(x) \frac{v(y', y_4 - |x' - y'|)}{|x' - y'|}\, dx\, dy'.$$

This gives the well known formula for the retarded potential:

$$E^+ * v(x) = \frac{1}{4\pi}\int \frac{v(y', x_4 - |x'-y'|)}{|x'-y'|}\, dy' \tag{7.3.15}$$

which also holds, for example, when v is bounded and measurable. Similarly, one can show that

$$E^- * v(x) = \frac{1}{4\pi}\int \frac{v(y', x_4 + |x'-y'|)}{|x'-y'|}\, dy', \tag{7.3.16}$$

when $v \in C_c^\infty(\mathbf{R}^4)$, for example. So one can also write (7.3.14) as

$$\langle E^+ * v, \phi\rangle = \langle v, E^- * \phi\rangle, \quad \phi \in C_c^\infty(\mathbf{R}^4). \tag{7.3.17}$$

Exercises

7.1. (i) Let f: $\mathbf{R}^n \to \mathbf{R}^{n+m}$ be given by $y = f(x)$, where

$y_i = x_i, \quad i = 1, \ldots, n, \quad y_i = 0, \quad i = h+1, \ldots, n+m.$

Let $\psi \in C_c(\mathbf{R}^{n+m})$ be such that $\int \psi \, dy = 1$, and set $\psi_j = j^{n+m}\psi(jy)$, $j = 1, 2, \ldots$, so that $\psi_j \to \delta$ in $\mathscr{D}'(\mathbf{R}^{n+m})$ as $j \to \infty$. Show that the $f^*\psi_j$ do not converge in $\mathscr{D}'(\mathbf{R}^n)$.

(ii) Show that, if f: $\mathbf{R} \to \mathbf{R}$ is $x \to y = x^m$, where $m > 1$ is an integer, then $f^*\delta$ cannot be defined as the limit, in $\mathscr{D}'(\mathbf{R})$, of the pullbacks of a sequence of test functions.

7.2. Prove that, if $f \in C^\infty(\mathbf{R}^n)$ and $f(x) = 0$ implies that $f'(x) \neq 0$, then division by f in $\mathscr{D}'(\mathbf{R}^n)$ can be reduced to the solutions of the division problem in $\mathscr{D}'(\mathbf{R})$.

7.3. For any $w \in C^\infty(\mathbf{R}^2)$, set $w^c = wH(x_4)$, where H is the Heaviside function. Let P be the d'Alembertian (7.3.7), and $u \in C^\infty(\mathbf{R}^4)$. Show that

$Pu^c = f^c + u_0(x') \otimes \partial\delta(x_4) + u_1(x') \otimes \delta(x_4),$

where $f = Pu$, $u_0 = u(x', 0)$, and $u_1 = \partial u(x', 0)$. Deduce that u^c is the sum of the retarded potential of f^c and of the Poisson representation of the solution of the initial value problem,

$$\frac{(x_4)_+}{4\pi}\int_{S^2} u_0(x' + x_4\theta)\, d\omega(\theta) + \partial_4\left(\frac{(x_4)_+}{4\pi}\int_{S^2} u_1(x' + x_4\theta)\, d\omega(\theta)\right).$$

7.4. Write $x \in \mathbf{R}^n$ as (x', x_n), where $x' \in R^{n-1}$, and let

$f(x) = x_n^2 - |x'|^2.$

Take $F \in \mathscr{D}'(\mathbf{R})$ such that supp $F \subset \bar{\mathbf{R}}^+$. Show that pullback of F by f: $\mathbf{R}^n \backslash \{0\} \to \mathbf{R}$ can be split up as

$f^*F = F_+(f) + F_-(f)$

where

$\operatorname{supp} F_+(f) \subset \{x : x_n \geqslant |x'|\}, \quad \operatorname{supp} F_-(f) \subset \{x : x_n \leqslant |x'|\}.$

Let

$P = \partial_n^2 - \partial_1^2 - \ldots - \partial_{n-1}^2$

be the d'Alembertian on $\mathbf{R}^n$. Show that, in $\mathscr{D}'(\mathbf{R}^n\backslash\{0\})$,

$$Pf^*F = 4f(f^*\partial^2F + 2nf^*\partial F).$$

Deduce that $PF_+(f) = 0$ on $\mathbf{R}^n\backslash\{0\}$ if

$$F = \delta^{(n-3/2)}(t) \quad \text{when } n \text{ is even,}$$
$$= t_+^{1-\frac{1}{2}n} \quad \text{when } n \text{ is odd.}$$

Now show that, with this F, $F_+(f)$ can be put into the form P^kG where $G \in L_1^{\text{loc}}(\mathbf{R}^n)$ if the integer k is taken large enough. Hence show that $F_+(f)$ extends to a member of $\mathscr{D}'(\mathbf{R}^n)$ which is homogeneous of degree $2-n$, and prove that the unique forward fundamental solution of the d'Alembertian on $\mathbf{R}^n$ is

$$E^+ = \frac{1}{2\pi^m}\,\delta_+^{(m-1)}(x_n^2 - |x'|^2) \quad \text{if } n = 2m+2, \quad n \geqslant 1$$

and

$$E^+ = \frac{H(x_n)}{2\pi^{m-\frac{1}{2}}\Gamma(\frac{3}{2}-m)}\,(x_n^2 - |x'|^2)_+^{\frac{1}{2}-m} \quad \text{if } n = 2m+1, \quad m \geqslant 1.$$

(In 'pre-distribution' form, these results are due to Hadamard and M. Riesz; for another derivation, and detailed references, see, for example, the author's *The wave equation on a curved space-time*, Cambridge University Press, 1975, Ch. 6.)

8 TEMPERED DISTRIBUTIONS AND FOURIER TRANSFORMS

8.1. Introduction

The Fourier transform of a function f on $\mathbf{R}^n$ is

$$\hat{f}(\xi) = \int f(x)e^{-ix\cdot\xi}\,dx, \quad \xi \in \mathbf{R}^n, \tag{8.1.1}$$

where $x \cdot \xi$ is the usual inner product on $\mathbf{R}^n$,

$$x \cdot \xi = \sum_{j=1}^{n} x_j \xi_j.$$

Some hypothesis on f is needed, to ensure the existence of the integral. A simple choice is to take $f \in L_1(\mathbf{R}^n)$, the (vector) space of equivalence classes of absolutely integrable functions. This is actually also a space of distributions; its members are the distributions which can be realized in the form

$$\phi \mapsto \int f\phi\,dx, \quad \phi \in C_c^\infty(\mathbf{R}^n),$$

where f is a (measurable) function such that

$$\|f\|_{L_1} = \int |f(x)|\,dx < \infty. \tag{8.1.2}$$

Clearly, $\|\ \|_{L_1}$ is a norm on this space; it is a basic result in integration theory that $L_1(\mathbf{R}^n)$ is complete with respect to this norm, and so is a Banach space.

The space $L_1(X)$, where X is an open set in $\mathbf{R}^n$, is defined in the same way. At this point, it is useful to recall Fubini's theorem, which will be stated without proof.

Theorem 8.1.1. Let $X \subset \mathbf{R}^n$ and $Y \subset \mathbf{R}^m$ be open sets, and let $f \in L_1(X \times Y)$. Then the functions

$$x \mapsto \int_Y f(x,y)\,dy, \quad y \mapsto \int_X f(x,y)\,dx$$

exist for almost all x and almost all y, respectively, and are integrable; furthermore,

$$\int_{X\times Y} f(x,y)\,\mathrm{d}x\,\mathrm{d}y = \int_X \mathrm{d}x \int_Y f(x,y)\,\mathrm{d}y = \int_Y \mathrm{d}y \int_X f(x,y)\,\mathrm{d}x. \quad (*)$$

Also, if any one of the integrals

$$\int_{X\times Y} |f(x,y)|\,\mathrm{d}x\,\mathrm{d}y, \quad \int_X \mathrm{d}x \int_Y |f(x,y)|\,\mathrm{d}y,$$

$$\int_Y \mathrm{d}y \int_X |f(x,y)|\,\mathrm{d}x$$

is finite, then so are the other two, and $(*)$ holds.

Fubini's theorem can be used to prove the following theorem on integrals depending on a parameter:

Theorem 8.1.2. Let $X\subset\mathbf{R}^n$ be an open set, and $J\subset\mathbf{R}$ an open interval. Let f be a function on $X\times J$ with the following properties:

(i) the partial derivative $\partial_t f = f_t'(x,t)$ exists, and is a continuous function, for almost all $x\in X$;

(ii) there is a nonnegative $g\in L_1(X)$ such that $|f_t'|\leqslant g$ for $t\in J$;

(iii) the integral $\int_X f(x,t_0)\,\mathrm{d}x$ exists for some $t_0\in J$. Then

$$F(t) = \int_X f(x,t)\,\mathrm{d}x \qquad (8.1.3)$$

exists for all $t\in J$ and is continuously differentiable, and

$$F'(t) = \int_X f_t'(x,t)\,\mathrm{d}x. \qquad (8.1.4)$$

Proof. By (ii), the second member of (8.1.4) exists for all $t\in J$; let us call it F_1,

$$F_1(t) = \int_X f_t'(x,t)\,\mathrm{d}x.$$

It also follows from (i), (ii) and dominated convergence, that F_1 is continuous on J. Now if $t\in J$ then

$$\int_{t_0}^{t} F_1(s)\,\mathrm{d}s = \int_{t_0}^{t}\mathrm{d}s \int_X f_t'(x,s)\,\mathrm{d}x = \int_X \mathrm{d}x \int_{t_0}^{t} f_t'(x,s)\,\mathrm{d}s$$

by (ii) and Fubini's theorem. But

$$\int_{t_0}^{t} f_t'(x,s)\,\mathrm{d}s = f(x,t) - f(x,t_0)$$

for almost all $x \in X$; so (iii) gives the identity

$$F(t) = \int_X f(x, t_0)\,\mathrm{d}x + \int_{t_0}^{t} F_1(s)\,\mathrm{d}s.$$

As F_1 is continuous, all the assertions follow from this, and we are done.

Theorem 8.12 extends in an obvious way to a function f defined on $X \times Y$, where $Y \subset \mathbf{R}^m$ is an open set.

The next theorem lists some elementary properties of the Fourier transform on $L_1(\mathbf{R}^n)$.

Theorem 8.1.3. (i) Let $f \in L_1(\mathbf{R}^n)$. Then its Fourier transform exists, is continuous, and bounded by the L_1-norm of f,

$$|\hat{f}(\xi)| \leqslant \|f\|_{L_1(\mathbf{R}^n)}, \quad \xi \in \mathbf{R}^n. \tag{8.1.5}$$

(ii) If $f \in L_1(\mathbf{R}^n)$ and $g \in L_1(\mathbf{R}^n)$, then

$$\int f(x)\hat{g}(x)\,\mathrm{d}x = \int \hat{f}(\xi)g(\xi)\,\mathrm{d}\xi. \tag{8.1.6}$$

(iii) If $f \in L_1(\mathbf{R}^n)$ and $g \in L_1(\mathbf{R}^n)$, then the convolution of f and g exists for almost all $x \in \mathbf{R}^n$, and is a member of $L_1(\mathbf{R}^n)$; moreover,

$$(f * g)\hat{} = \hat{f}\hat{g}. \tag{8.1.7}$$

Proof. (i) This follows from

$$|f(x)\mathrm{e}^{-\mathrm{i}x\cdot\xi}| = |f(x)|, \quad \xi \in \mathbf{R}^n,$$

and the continuity of $\hat{f}$ is implied by the dominated convergence theorem.

(ii) By (i), one has $f\hat{g} \in L_1(\mathbf{R}^n)$, and clearly

$$\int f(x)\hat{g}(x)\,\mathrm{d}x = \int f(x)\,\mathrm{d}x \int g(\xi)\mathrm{e}^{-\mathrm{i}x\cdot\xi}\,\mathrm{d}\xi = \int g(\xi)\,\mathrm{d}\xi \int f(x)\mathrm{e}^{-\mathrm{i}x\cdot\xi}\,\mathrm{d}x$$

by Fubini's theorem; this gives (8.1.6).

(iii) One first notes that

$$\int |f(y)g(x-y)|\,\mathrm{d}x\,\mathrm{d}y = \|f\|_{L_1(\mathbf{R}^n)}\|g\|_{L_1(\mathbf{R}^n)} < \infty.$$

(It can be shown that $f(y)g(x-y)$ is measurable.) So Fubini's theorem (Theorem 8.1.1) shows that $f * g$ exists for almost all x, and is in $L_1(\mathbf{R}^n)$. Another appeal to Fubini's theorem gives then

$$\begin{aligned}\int (f*g)\hat{} &= \int \mathrm{e}^{-\mathrm{i}x\cdot\xi}\,\mathrm{d}x \int f(y)g(x-y)\,\mathrm{d}y \\ &= \int f(y)\,\mathrm{d}y \int g(x-y)\mathrm{e}^{-\mathrm{i}x\cdot\xi}\,\mathrm{d}x\end{aligned}$$

and this yields (8.1.7) if one puts $x = y + z$ in the inner integral,

$$(f * g)\hat{} = \int f(y)\, \mathrm{d}y \int g(z) \mathrm{e}^{-\mathrm{i}x\cdot\xi} \mathrm{e}^{-\mathrm{i}y\cdot\xi}\, \mathrm{d}z = \hat{f}\hat{g}.$$

This completes the proof of the theorem.

In addition to (8.1.5), one can also prove that $\hat{f}(\xi) \to 0$ as $|\xi| \to \infty$; this is the Riemann-Lebesgue lemma.

If $f \in L_1(\mathbf{R}^n)$, then the identity (8.1.6), with $g \in C_c^\infty(\mathbf{R}^n)$, gives the distribution determined by $\hat{f}$ in terms of f. But one cannot use duality to extend this to $f \in \mathscr{D}'(\mathbf{R}^n)$, because the Fourier transform is not a map $C_c^\infty(\mathbf{R}^n) \to C_c^\infty(\mathbf{R}^n)$. For the Fourier transform of a test function extends to an analytic function on $\mathbf{C}^n$. The restriction of such a function to the reals cannot have compact support unless it is identically zero, by the uniqueness of analytic continuation. (See Chapter 10.) L. Schwartz observed that this difficulty can be overcome by introducing a new space of test functions, whose dual is a subspace of $\mathscr{D}'(\mathbf{R}^n)$. We shall now go on to develop this theory.

From now on, we shall use the operators

$$D_j = \frac{1}{\mathrm{i}} \partial_j, \quad j = 1, \ldots, n \quad (\mathrm{i} = \sqrt{-1}) \tag{8.1.8}$$

instead of the ∂_j. This simplifies the notation, as the Fourier transform maps the operation D_j to multiplication by ξ_j.

8.2. Rapidly decreasing test functions

We begin by introducing a new space of test functions.

Definition 8.2.1. A function $\phi \in C^\infty(\mathbf{R}^n)$ is called rapidly decreasing if

$$\|\phi\|_{\alpha,\beta} = \sup |x^\alpha D^\beta \phi| < \infty \tag{8.2.1}$$

for all pairs of multi-indices α, β; the vector space of such functions is called $\mathscr{S}(\mathbf{R}^n)$. A sequence $(\phi_j)_{1 \leqslant j < \infty} \in \mathscr{S}(\mathbf{R}^n)$ is said to converge to zero in $\mathscr{S}(\mathbf{R}^n)$ if $\|\phi_j\|_{\alpha,\beta} \to 0$ as $j \to \infty$ for all α and β.

The reason for the terminology is that, if $\phi \in \mathscr{S}(\mathbf{R}^n)$, then each $D^\alpha \phi$ tends to 0 faster than $|x|^{-N}$ for all $N \geqslant 0$ as $|x| \to \infty$. An example of a rapidly decreasing function is $x \mapsto \exp(-c|x|^2)$, where $\operatorname{Re} c > 0$.

The semi-norms (8.2.1) generate a topology, and it is an easy exercise to show that $\mathscr{S}(\mathbf{R}^n)$ is complete when it is equipped with this topology. Hence $\mathscr{S}(\mathbf{R}^n)$ is a Fréchet space. Note that, in particular, sequential continuity is equivalent to continuity.

There are some immediate consequences of the definition which we collect as a theorem.

Theorem 8.2.1. (i) $\mathscr{S}(\mathbf{R}^n)$ is stable under differentiation, and under multiplication by polynomials; moreover, if p and q are polynomials, then $\phi \mapsto p(x)q(D)\phi$ is a continuous map $\mathscr{S}(\mathbf{R}^n) \to \mathscr{S}(\mathbf{R}^n)$.

(ii) $C_c^\infty(\mathbf{R}^n) \subset \mathscr{S}(\mathbf{R}^n)$, and the injection is continuous.

(iii) $C_c^\infty(\mathbf{R}^n)$ is dense in $\mathscr{S}(\mathbf{R}^n)$.

(iv) $\mathscr{S}(\mathbf{R}^n) \subset L_1(\mathbf{R}^n)$, and the injection is continuous.

Proof. Property (i) is obvious from Definition 8.2.1, and so is (ii), since $\phi \to 0$ on $C_c^\infty(\mathbf{R}^n)$ clearly implies that also $\phi \to 0$ in $\mathscr{S}(\mathbf{R}^n)$. To prove (iii), let $\phi \in \mathscr{S}(\mathbf{R}^n)$ be given; choose some $\rho \in C_c^\infty(\mathbf{R}^n)$ such that $\rho = 1$ for $|x| \leqslant 1$, and set $\phi_j = \rho(x/j)\phi(x)$ where $j = 1, 2, \ldots$. Then there are constants $C_N, N = 1, 2, \ldots$ such that

$$\sum_{|\alpha| \leqslant N, |\beta| \leqslant N} \|\phi - \phi_j\|_{\alpha,\beta} \leqslant C_N \sum_{|\alpha| \leqslant N, |\beta| \leqslant N} \sup\{|x^\alpha D^\beta \phi(x)| : |x| \geqslant j\}.$$

The right hand side converges to 0 as $j \to \infty$, since it follows from (8.2.1) that

$$|x^\alpha D^\beta \phi(x)| \leqslant |x|^{-1} \sum_{|\gamma| = |\alpha| + 1} \|\phi\|_{\gamma,\beta}.$$

Finally, let N be an integer such that $N > \frac{1}{2}n$, so that $(1 + |x|^2)^{-N} \in L_1(\mathbf{R}^n)$. Let $\phi \in \mathscr{S}(\mathbf{R}^n)$; then

$$(1 + |x|^2)^N |\phi(x)| \leqslant C \sum_{|\alpha| \leqslant 2N} \|\phi\|_{\alpha,0}$$

for some constant C, by (8.2.1). Hence, with a different C,

$$\|\phi\|_{L_1(\mathbf{R}^n)} \leqslant C \sum_{|\alpha| \leqslant 2N} \|\phi\|_{\alpha,0} \tag{8.2.2}$$

and this proves (iv).

We shall now show that the Fourier transform is a continuous isomorphism of $\mathscr{S}(\mathbf{R}^n)$. We proceed in steps.

Lemma 8.2.1. If $\phi \in \mathscr{S}(\mathbf{R}^n)$, then

$$(D^\alpha \phi)\hat{} = \xi^\alpha \hat{\phi}(\xi) \tag{8.2.3}$$

and

$$(x^\alpha \phi)\hat{} = (-1)^{|\alpha|} D^\alpha \hat{\phi}(\xi). \tag{8.2.4}$$

Proof. These Fourier transforms exist, by parts (i) and (iv) of Theorem 8.2.1. The identity (8.2.3) is proved by integration by parts. Formally, (8.3.4) results from differentiation under the integral sign, and the reader should have no difficulty in verifying that this can be justified by means of Theorem 8.1.2.

Lemma 8.2.2. The Fourier transform is a continuous map $\mathscr{F}:\mathscr{S}(\mathbf{R}^n)\to\mathscr{S}(\mathbf{R}^n)$.

Proof. It follows from (8.2.4) and Theorem 8.2.1 (iv) that, for any multi-index α, $D^\alpha\phi$ is the Fourier transform of a function of class L_1 if $\phi\in\mathscr{S}(\mathbf{R}^n)$; hence it is continuous by Theorem 8.1.3, and so $\hat{\phi}\in C^\infty(\mathbf{R}^n)$. Again, (8.2.3), (8.2.4) and (8.1.5) give

$$\|\hat{\phi}\|_{\alpha,\beta} = \|(-1)^{|\beta|}(D^\alpha(x^\beta\phi))\hat{}\,\| \leqslant \|D^\alpha(x^\beta\phi)\|_{L_1(\mathbf{R}^n)},$$

and the lemma now follows from parts (i) and (iv) of Theorem 8.2.1, or directly from (8.2.2) and Leibniz's theorem.

Lemma 8.2.3. In $\mathbf{R}^n$, one has

$$(\exp(-\tfrac{1}{2}|x|^2))\hat{} = (2\pi)^{\frac{1}{2}n}\exp(-\tfrac{1}{2}|\xi|^2). \tag{8.2.5}$$

Proof. This well known result, usually proved by contour integration, can be deduced from Lemma 8.2.1. If $n=1$, then $\phi(x)=\exp(-\frac{1}{2}x^2)$ satisfies the differential equation

$$iD\phi + x\phi = 0.$$

So (8.2.3) and (8.2.4) give

$$iD\hat{\phi} + \xi\hat{\phi} = 0,$$

whence $\hat{\phi}=C\exp(-\frac{1}{2}\xi^2)$. To determine C we have

$$C = \hat{\phi}(0) = \int e^{-\frac{1}{2}x^2}\,dx = (2\pi)^{\frac{1}{2}}.$$

Finally,

$$(\exp(-\tfrac{1}{2}|x|^2))\hat{} = \prod_{j=1}^{n}\int e^{-ix_j\xi_j-\frac{1}{2}x_j^2}\,dx_j = \prod_{j=1}^{n}\hat{\phi}(\xi_j),$$

and this is (8.2.5).

The last step is the inversion theorem ('Fourier's theorem'):

Theorem 8.2.2. If $\phi\in\mathscr{S}(\mathbf{R}^n)$ then

$$\phi(x) = (2\pi)^{-n}\int\hat{\phi}(\xi)e^{ix\cdot\xi}\,d\xi. \tag{8.2.6}$$

Proof. It follows from (8.1.6) that

$$\int\hat{\phi}(\xi)\psi(\xi)\,d\xi = \int\phi(x)\hat{\psi}(x)\,dx \tag{8.2.7}$$

if $\phi\in\mathscr{S}(\mathbf{R}^n)$ and $\psi\in\mathscr{S}(\mathbf{R}^n)$. Let ϵ be a positive real number, and take $\psi=\exp(-\frac{1}{2}\epsilon^2|\xi|^2)$. Then it follows from Lemma 8.2.3 that (8.2.7) becomes

$$\int \hat{\phi}(\xi) \exp\left(-\tfrac{1}{2}\epsilon^2 |\xi|^2\right) d\xi = (2\pi)^{\frac{1}{2}n} \epsilon^{-n} \int \phi(x) \exp\left(-\frac{1}{2\epsilon^2}|x|^2\right) dx$$

$$= (2\pi)^{\frac{1}{2}n} \int \phi(\epsilon z) \exp\left(-\tfrac{1}{2}|z|^2\right) dz.$$

Now $\hat{\phi} \in \mathscr{S}(\mathbf{R}^n) \subset L_1(\mathbf{R}^n)$ and $0 < \exp(-\frac{1}{2}\epsilon^2 |\xi|^2) \leqslant 1$, so one can let $\epsilon \to 0$ under the integral sign on the left hand side, by dominated convergence. Again, $\exp(-\frac{1}{2}|z|^2) \in L_1(\mathbf{R}^n)$, and $\phi(\epsilon z)$ is bounded; so one can also let $\epsilon \to 0$ under the integral sign on the right hand side, by dominated convergence. So

$$\int \hat{\phi}(\xi)\, d\xi = (2\pi)^{\frac{1}{2}n} \phi(0) \int \exp\left(-\tfrac{1}{2}|x|^2\right) dx,$$

whence

$$\phi(0) = (2\pi)^{-n} \int \hat{\phi}(\xi)\, d\xi. \tag{8.2.8}$$

A simple change of variable of integration in (8.1.1) shows that

$$(\tau_{-h}\phi)\hat{} = \hat{\phi}(e) e^{i\xi \cdot h}, \quad h \in \mathbf{R}^n.$$

Hence (8.2.8), applied to $\tau_{-x}\phi$, gives (8.2.6), and we are done.

One can now combine Lemma 8.2.2 and Theorem 8.2.2, to obtain

Theorem 8.2.3. The Fourier transform is a continuous isomorphism $\mathscr{F}: \mathscr{S}(\mathbf{R}^n) \to \mathscr{S}(\mathbf{R}^n)$, and its inverse is a continuous isomorphism $\mathscr{F}^{-1}: \mathscr{S}(\mathbf{R}^n) \to \mathscr{S}(\mathbf{R}^n)$.

Proof. If $\phi \in \mathscr{S}(\mathbf{R}^n)$ and $\hat{\phi} = 0$, then $\phi = 0$ by (8.2.6); so the map $\mathscr{F}$ is injective. Again, the inversion formula (8.2.6) can be written as

$$\phi(x) = (2\pi)^{-n} \int \hat{\phi}(-\xi) e^{-ix\cdot\xi}\, d\xi = \hat{\psi}(x) \tag{8.2.9}$$

where $\psi(\xi) = (2\pi)^{-n} \hat{\phi}(-\xi) \in \mathscr{S}(\mathbf{R}^n)$. Hence the Fourier transform map $\mathscr{F}$ is also surjective, and so an isomorphism. The continuity of $\mathscr{F}$ has already been proved (Lemma 8.2.2); as the reflection map $\phi(x) \mapsto \phi(-x)$ is evidently also a continuous map $\mathscr{S}(\mathbf{R}^n) \to \mathscr{S}(\mathbf{R}^n)$, it follows from (8.2.9) that $\mathscr{F}^{-1}$ is continuous, and the theorem is proved.

Note that one can write (8.2.9) more concisely, as

$$\hat{\hat{\phi}} = (2\pi)^n \check{\phi}, \quad \phi \in \mathscr{S}(\mathbf{R}^n). \tag{8.2.10}$$

8.3. Tempered distributions

The dual of $\mathscr{S}(\mathbf{R}^n)$ consists of continuous linear forms on $\mathscr{S}(\mathbf{R}^n)$; obviously, such a linear form u is continuous if and only if there is a constant

$C \geqslant 0$ and a nonnegative integer N such that

$$|\langle u, \phi\rangle| \leqslant C \sum_{|\alpha|, |\beta| \leqslant N} \sup |x^\alpha D^\beta \phi|, \quad \phi \in \mathscr{S}(\mathbf{R}^n). \tag{8.3.1}$$

Arguing as in the proof of Theorem 1.3.2, one can show that a linear form u on $\mathscr{S}(\mathbf{R}^n)$ is continuous if and only if it is sequentially continuous, that is, if $\phi_j \to 0$ in $\mathscr{S}(\mathbf{R}^n)$ implies that $\langle u, \phi_j\rangle \to 0$ as $j \to \infty$; this is left as an exercise.

By part (ii) of Theorem 8.2.1, the restriction of a continuous linear form u on $\mathscr{S}(\mathbf{R}^n)$ to $C_c^\infty(\mathbf{R}^n)$ is a distribution; by part (iii) of the same theorem, u is uniquely determined by its restriction to $C_c^\infty(\mathbf{R}^n)$. We therefore make the following definition:

Definition 8.3.1. $\mathscr{S}'(\mathbf{R}^n)$ is the subspace of $\mathscr{D}'(\mathbf{R}^n)$ consisting of distributions which extend to continuous linear forms on $\mathscr{S}(\mathbf{R}^n)$.

A sequence $(u_j)_{1 \leqslant j < \infty} \in \mathscr{S}'(\mathbf{R}^n)$ is said to converge to $u \in \mathscr{S}'(\mathbf{R}^n)$ in $\mathscr{S}'(\mathbf{R}^n)$ if $\langle u_j, \phi\rangle \to \langle u, \phi\rangle$ for all $\phi \in \mathscr{S}(\mathbf{R}^n)$ as $j \to \infty$.

The members of $\mathscr{S}'(\mathbf{R}^n)$ are called *tempered* (or, temperate) *distributions.* The space of these is stable under differentiation, and multiplication by polynomials; this is the dual of part (i) of Theorem 8.2.1.

One clearly has both $L_1(\mathbf{R}^n)$ and $\mathscr{E}'(\mathbf{R}^n) \subset \mathscr{S}'(\mathbf{R}^n)$. A more telling example of a tempered distribution is a *continuous function of polynomial growth.* By definition, if f is such a function, then there are constants $C, M \geqslant 0$ such that

$$|f(x)| \leqslant C(1 + |x|)^M, \quad x \in \mathbf{R}^n. \tag{8.3.2}$$

By the remark made above, the derivatives of such a function are also tempered distributions. The *structure theorem* for $\mathscr{S}'(\mathbf{R}^n)$ asserts that all tempered distributions can be represented in this fashion:

Theorem 8.3.1. Every tempered distribution is a derivative of finite order of some continuous function of polynomial growth.

Proof. It is sufficient to prove this for a tempered distribution supported in the octant $\Omega = \{x : x_1 > 0, \ldots, x_n > 0\}$. For this implies the result when $u \in \mathscr{S}'(\mathbf{R}^n)$ is supported in $\{x: \epsilon_1 x_1 > -\delta, \ldots, \epsilon_n x_n > -\delta\}$, where $\delta > 0$ and $\epsilon_j = 1$ or $\epsilon_j = -1$, and, via a partition of unity, any tempered distribution can be written as the sum of 2^n distributions of this type.

Let us therefore take $u \in \mathscr{S}'(\mathbf{R}^n)$ with $\operatorname{supp} u \subset \Omega$. Then one has a semi-norm estimate (8.3.1). Take any $\phi \in C_c^\infty(\mathbf{R}^n)$, and choose some fixed $\rho \in C^\infty(\mathbf{R}^n)$ which is supported in Ω and such that $\rho = 1$ on a neighbourhood of $\operatorname{supp} u$. Then (8.3.1), applied to $\rho\phi$, gives, with a different constant C,

$$|\langle u,\phi\rangle|\leqslant C\sum_{|\alpha|,|\beta|\leqslant N}\sup\{x^\alpha|D^\beta\phi(x)|: x\in\Omega\},\quad \phi\in C_c^\infty(\mathbf{R}^n),\quad (8.3.3)$$

since $|x^\alpha|=x^\alpha$ when $x\in\Omega$.

Let

$$E_{N+2}=\frac{(x_1)_+^{N+1}\dots(x_n)_+^{N+1}}{((N+1)!)^n}.$$

By Theorem 5.3.1, $E_{N+2}*u$ is well defined, is supported in Ω, and

$$u=(\partial_1\dots\partial_n)^{N+2}(E_{N+2}*u).$$

As in the proof of Theorem 5.4.1, one can now show, by regularizing E_{N+2} and a passage to the limit, that

$$E_{N+2}*u=\langle u(t),E_{N+2}(x-t)\rangle\in C^0(\mathbf{R}^n),$$

and that (8.3.3) can be applied to this. Hence, with $C'=C/((N+1)!)^n$,

$$|E_{N+2}*u(x)|$$
$$\leqslant C'\sum_{|\alpha|,|\beta|\leqslant N}\sup\{t^\alpha D^\beta(x_1-t_1)_+^{N+1}\dots(x_n-t_n)_+^{N+1}: t\in\Omega\}.$$

An elementary computation, which is left to the reader, shows that this implies that the (continuous) function $E_{N+2}*u$ satisfies an estimate of type (8.3.2), with $M=2N+1$. So the theorem is proved.

We shall now define the Fourier transform on $\mathscr{S}'(\mathbf{R}^n)$. If $u\in L_1(\mathbf{R}^n)$, then its 'classical' Fourier transform $\hat{u}(\xi)$ is a bounded and continuous function (Theorem 8.1.3 (i)), hence is in $\mathscr{S}'(\mathbf{R}^n)$ as a distribution. If now $\phi\in\mathscr{S}(\mathbf{R}^n)$ then both ϕ and $\hat{\phi}$ are in $L_1(\mathbf{R}^n)$ and so, by Theorem 8.1.3 (ii)),

$$\int\hat{u}(\xi)\phi(\xi)\,d\xi=\int u(x)\hat{\phi}(x)\,dx,\quad u\in L_1(\mathbf{R}^n),\quad \phi\in\mathscr{S}(\mathbf{R}^n),\qquad (8.3.4)$$

which one can write as $\langle\hat{u},\phi\rangle=\langle u,\hat{\phi}\rangle$. As $\phi\mapsto\hat{\phi}$ is a continuous map $\mathscr{S}(\mathbf{R}^n)\to\mathscr{S}(\mathbf{R}^n)$ (Lemma 8.2.2), $\phi\mapsto\langle u,\hat{\phi}\rangle$ is a tempered distribution whenever u is one. Guided by these observations, we make the following definition:

Definition 8.3.2. The Fourier transform of $u\in\mathscr{S}'(\mathbf{R}^n)$ is the distribution $\hat{u}\in\mathscr{S}'(\mathbf{R}^n)$ given by

$$\langle\hat{u},\phi\rangle=\langle u,\hat{\phi}\rangle,\quad \phi\in\mathscr{S}(\mathbf{R}^n).\qquad (8.3.5)$$

Remark. It follows from this definition, (8.3.4), and (i) of Theorem 8.1.3, that if $u\in L_1(\mathbf{R}^n)$, then its Fourier transform when it is regarded as a distribution is equal to its usual Fourier transform.

The basic properties of the Fourier transform of tempered distributions can now easily be deduced by duality from Theorem 8.2.3.

Theorem 8.3.2. The Fourier transform is an isomorphism $\mathcal{F}:\mathcal{S}'(\mathbf{R}^n)\to\mathcal{S}'(\mathbf{R}^n)$; both $\mathcal{F}$ and its inverse $\mathcal{F}^{-1}$ are sequentially continuous maps $\mathcal{S}'(\mathbf{R}^n)\to\mathcal{S}'(\mathbf{R}^n)$.

Proof. That $\mathcal{F}$ is linear is obvious. It follows from (8.3.5) and (8.2.10) that, if $\phi\in\mathcal{S}(\mathbf{R}^n)$, then

$$\langle\hat{\hat{u}},\phi\rangle=\langle\hat{u},\hat{\phi}\rangle=\langle u,\hat{\hat{\phi}}\rangle=(2\pi)^n\langle u,\check{\phi}\rangle.$$

Hence

$$\langle u,\check{\phi}\rangle=(2\pi)^{-n}\langle\hat{\hat{u}},\phi\rangle,\qquad \phi\in\ \ (\mathbf{R}^n). \tag{8.3.6}$$

(This is the inversion theorem for the Fourier transform on $\mathcal{S}'(\mathbf{R}^n)$.) So one has $u=0$ if $\hat{u}=0$, and $u\in\mathcal{S}'(\mathbf{R}^n)$ is the Fourier transform of $(2\pi)^{-n}(\hat{u})^{\check{}}\in\mathcal{S}'(\mathbf{R}^n)$, which means that $\mathcal{F}$ is both injective and surjective, hence an isomorphism. The sequential continuity of $\mathcal{F}$ follows from the definition, (8.3.5), that of $\mathcal{F}^{-1}$ from (8.3.6), and so we are done.

The sequential continuity of $\mathcal{F}$ also has the following consequence:

Corollary 8.3.1. The Fourier transform on $\mathcal{S}'(\mathbf{R}^n)$ is the extension of the form $\phi\mapsto\langle\hat{f},\phi\rangle$, where $f\in\mathcal{S}(\mathbf{R}^n)$ and $\hat{f}$ is its classical Fourier transform (8.3.1).

We now list some operational rules:

$$(D^\alpha u)\hat{}=\xi^\alpha\hat{u}, \tag{8.3.7}$$

$$(x^\alpha u)\hat{}=(-1)^{|\alpha|}D^\alpha\hat{u}, \tag{8.3.8}$$

$$(\tau_h u)\hat{}(\xi)=\hat{u}(\xi)e^{-i\xi\cdot h},\quad h\in\mathbf{R}^n, \tag{8.3.9}$$

$$(ue^{ix\cdot h})\hat{}=\tau_h\hat{u},\quad h\in\mathbf{R}^n. \tag{8.3.10}$$

All these are obtained by duality, and simple manipulations. For instance, one has

$$\begin{aligned}\langle(D^\alpha u)\hat{},\phi\rangle&=\langle D^\alpha u,\hat{\phi}\rangle=(-1)^{|\alpha|}\langle u,D^\alpha\hat{\phi}\rangle\\&=\langle u,(x^\alpha\phi)\hat{}\rangle\end{aligned}$$

by (8.2.4), and this gives (8.3.7). Again,

$$\langle(\tau_h u)\hat{},\phi\rangle=\langle\tau_h u,\hat{\phi}\rangle=\langle u,\tau_{-h}\hat{\phi}\rangle.$$

Now

$$\tau_{-h}\hat{\phi}(\xi)=\phi(\xi+h)=\int e^{-ix\cdot(\xi+h)}\phi(x)\,d\xi=(\phi e^{-ix\cdot h})\hat{}$$

hence

$$\langle(\tau_h u)\hat{},\phi\rangle=\langle u,(\phi e^{-ix\cdot h})\hat{}\rangle=\langle\hat{u}(\xi),\phi(\xi)e^{-i\xi\cdot h}\rangle$$

which is (8.3.9). The proof of (8.3.8) and (8.3.10) is left as an exercise.

Alternatively, one can use Corollary 8.3.1. To illustrate this, let us compute $(A^*u)\hat{}$, where $u\in\mathcal{S}'(\mathbf{R}^n)$, and A is an invertible (constant) $n\times n$ matrix. Now

A^*u, as in Definition 4.2.2, was obtained by duality, but is readily seen to be the extension of the form

$$\phi \mapsto \int u_j(Ax)\phi(x)\, dx, \quad \phi \in \mathcal{S}(\mathbf{R}^n),$$

when $u_j \in \mathcal{S}(\mathbf{R}^n)$ converges to u in $\mathcal{S}'(\mathbf{R}^n)$ as $j \to \infty$. But when $u \in \mathcal{S}(\mathbf{R}^n)$, one has

$$(A^*u)\hat{}(\xi) = \int e^{-ix\cdot\xi}u(Ax)\, dx = |\det A|^{-1} \int e^{-i(A^{-1}y)\cdot\xi}u(y)\, dy.$$

Now $(A^{-1}y)\cdot\xi = y\cdot({}^tA^{-1}\xi)$, where ${}^tA$ is the transpose of A, so

$$(A^*u)\hat{} = |\det A|^{-1}({}^tA^{-1})^*\hat{u}, \tag{8.3.11}$$

and this carries over to $\mathcal{S}'(\mathbf{R}^n)$ by continuity.

In particular, if $A = -I$, where I is the identity, then $A^*u = \check{u}$, and so one has

$$(\check{u})\hat{} = (\hat{u})\check{}. \tag{8.3.12}$$

Again, for a dilatation $A = tI$, where $t \in \mathbf{R}^+$, we get, if we write u_t for A^*u,

$$(u_t)\hat{} = t^{-n}\hat{u}_{1/t} \tag{8.3.13}$$

The simplest example is the Fourier transform of the Dirac distribution on $\mathbf{R}^n$. By (8.3.5),

$$\langle\hat{\delta}, \phi\rangle = \langle\delta, \hat{\phi}\rangle = \hat{\phi}(0) = \langle 1, \phi\rangle, \quad \phi \in \mathcal{S}(\mathbf{R}^n),$$

so that

$$\hat{\delta} = 1. \tag{8.3.14}$$

As $\check{\delta} = \delta$, the inversion formula (8.3.6) now gives

$$\hat{1} = (2\pi)^n\delta \tag{8.3.15}$$

which is nothing other than (8.2.8). Applying the operational rules (8.3.7)-(8.3.10) to (8.3.14) and (8.3.15) one gets

$$(D^\alpha\delta)\hat{} = \xi^\alpha, \quad (x^\alpha)\hat{} = (-1)^{|\alpha|}(2\pi)^n D^\alpha\delta,$$
$$(\tau_h\delta)\hat{} = e^{-ih\cdot\xi}, \quad (e^{ix\cdot h})\hat{} = (2\pi)^n\tau_h\delta.$$

Some other simple examples can be derived when $n = 1$, by starting with the signum function

$$\operatorname{sign} x = 1 \quad (x > 0), \quad \operatorname{sign} x = -1 \quad (x < 0), \quad \operatorname{sign} 0 = 0. \tag{8.3.16}$$

Clearly, $\partial(\operatorname{sign} x) = 2\delta$, so

$$i\xi(\operatorname{sign} x)\hat{} = 2$$

by (8.3.7) and (8.3.14). Hence

$$(\operatorname{sign} x)\hat{} = \frac{2}{i\xi} + C\delta(\xi),$$

where $1/\xi$ is the principal value distribution, and C is a constant. But sign x is odd (both as a function and as a distribution), so its Fourier transform is odd, by (8.3.12); hence $C=0$, and we have proved that

$$(\operatorname{sign} x)\hat{} = \frac{2}{\mathrm{i}\xi}. \tag{8.3.17}$$

As $\frac{1}{2}(1+\operatorname{sign} x) = H(x)$, the Heaviside function, one obtains from (8.3.17) and (8.3.15) with $n=1$ that

$$\hat{H} = \frac{1}{\mathrm{i}\xi} + \pi\delta. \tag{8.3.18}$$

Finally, the inversion formula (8.3.6), applied to (8.3.17), yields

$$\left(\frac{1}{x}\right)\hat{} = -\mathrm{i}\pi \operatorname{sign} \xi. \tag{8.3.19}$$

8.4. The convolution theorem

We have still to discuss the distribution form of part (iii) of Theorem 8.1.3, which is usually called the convolution theorem, and is one of the most important properties of the Fourier transform. It will be recalled that $u * v$ is well defined in $\mathscr{D}'$ if one of the factors has compact support; this is the case we shall consider, with the other factor a tempered distribution. We need a preliminary result which is of considerable intrinsic interest.

Theorem 8.4.1. If $u \in \mathscr{E}'(\mathbf{R}^n)$, then

$$\hat{u} = \langle u(x), \mathrm{e}^{-\mathrm{i}x\cdot\xi}\rangle, \tag{8.4.1}$$

and this is in $C^\infty(\mathbf{R}^n)$.

Proof. The function $\xi \mapsto \langle u(x), \exp(-\mathrm{i}x\cdot\xi)\rangle$ is C^∞, by Corollary 4.1.2. It has already been noted that $\mathscr{E}'(\mathbf{R}^n)$ is a subspace of $\mathscr{S}'(\mathbf{R}^n)$, so u certainly has a Fourier transform $\hat{u} \in \mathscr{S}'(\mathbf{R}^n)$, and this is uniquely determined by its restriction to $C_c^\infty(\mathbf{R}^n)$. Now if $\phi \in C_c^\infty(\mathbf{R}^n)$ then

$$\begin{aligned}\langle u(x)\otimes\phi(\xi), \mathrm{e}^{-\mathrm{i}x\cdot\xi}\rangle &= \left\langle u(x), \int \phi(\xi)\mathrm{e}^{-\mathrm{i}x\cdot\xi}\,\mathrm{d}\xi\right\rangle \\ &= \int \langle u(x), \mathrm{e}^{-\mathrm{i}x\cdot\xi}\rangle\phi(\xi)\,\mathrm{d}\xi,\end{aligned}$$

by the obvious variant of Theorem 4.3.6 for distributions with compact support. Hence

$$\langle\hat{u}, \phi\rangle = \langle u, \hat{\phi}\rangle = \int \langle u(x), \mathrm{e}^{-\mathrm{i}x\cdot\xi}\rangle\phi(\xi)\,\mathrm{d}\xi, \quad \phi \in C_c^\infty(\mathbf{R}^n)$$

and this proves (8.4.1).

The convolution theorem for distributions with compact support is an immediate consequence:

Corollary 8.4.1. If $u \in \mathscr{E}'(\mathbf{R}^n)$ and $v \in \mathscr{E}'(\mathbf{R}^n)$, then

$$(u * v)\hat{} = \hat{u}\hat{v}. \tag{8.4.2}$$

Proof. As $u * v \in \mathscr{E}'(\mathbf{R}^n)$, one can compute as follows:

$$(u * v)\hat{} = \langle u * v, \mathrm{e}^{-\mathrm{i}x\cdot\xi}\rangle$$
$$= \langle u(x) \otimes v(y), \mathrm{e}^{-\mathrm{i}(x+y)\cdot\xi}\rangle = \langle u(x), \mathrm{e}^{-\mathrm{i}x\cdot\xi}\rangle\, \langle v(y), \mathrm{e}^{-\mathrm{i}y\cdot\xi}\rangle$$

It is clear that (8.4.2) cannot hold in general unless $\hat{u}\hat{v}$ is well defined, and is a tempered distribution. So, if it is to hold for all $u \in \mathscr{S}'(\mathbf{R}^n)$, then $\hat{v}$ must be a *multiplier* on $\mathscr{S}'$. This class is defined as follows:

Definition 8.4.1. The class $\mathcal{O}_M(\mathbf{R}^n)$ of multipliers on $\mathscr{S}'(\mathbf{R}^n)$ consists of all $f \in C^\infty(\mathbf{R}^n)$ such that $D^\alpha f$ is of polynomial growth for all multi-indices α.

As an example, one can take $f = \exp(\mathrm{i}|x|^2)$. To justify this definition, it is sufficient to observe that, if $f \in \mathcal{O}_M(\mathbf{R}^n)$, then $\phi \mapsto f\phi$ is a continuous map $\mathscr{S}(\mathbf{R}^n) \to \mathscr{S}(\mathbf{R}^n)$; the easy proof is left as an exercise. So $u \mapsto fu$ is a homomorphism $\mathscr{S}'(\mathbf{R}^n) \to \mathscr{S}'(\mathbf{R}^n)$, by duality; it is also immediate that this is sequentially continuous.

Lemma 8.4.1. If $v \in \mathscr{E}'(\mathbf{R}^n)$, then $\hat{v} \in \mathcal{O}_M(\mathbf{R}^n)$.

Proof. It follows from Theorems 8.4.1 and 4.1.1 that

$$D^\alpha \hat{v}(\xi) = (-1)^{|\alpha|}\langle x^\alpha v(x), \mathrm{e}^{-\mathrm{i}x\cdot\xi}\rangle.$$

As $x^\alpha v$ is also in $\mathscr{E}'(\mathbf{R}^n)$, one thus has a semi-norm estimate

$$|D^\alpha \hat{v}(\xi)| \leqslant C_\alpha \sum_{|\beta| \leqslant N_\alpha} \sup\{|D^\beta(x^\alpha \mathrm{e}^{-\mathrm{i}x\cdot\xi})| : x \in K\}$$

where $K \subset \mathbf{R}^n$ is a compact set, and $C_\alpha \geqslant 0, N_\alpha \geqslant 0$. Hence there is a constant C'_α, for each α, such that

$$|D^\alpha \hat{v}(\xi)| \leqslant C'_\alpha(1 + |\xi|)^{N_\alpha},$$

and this proves the lemma.

We can now establish a version of the convolution theorem for distributions.

Theorem 8.4.2. Let $u \in \mathscr{S}'(\mathbf{R}^n)$ and $v \in \mathscr{E}'(\mathbf{R}^n)$. Then $u * v \in \mathscr{S}'(\mathbf{R}^n)$, and

$$(u * v)\hat{} = \hat{u}\hat{v}. \tag{8.4.3}$$

Proof. By Lemma 8.4.1, one has $\hat{u}\hat{v} \in \mathscr{S}'(\mathbf{R}^n)$. So there is a $w \in \mathscr{S}'(\mathbf{R}^n)$ such that $\hat{w} = \hat{u}\hat{v}$. To compute w, we use (8.3.6):

$$\langle w, \check{\phi} \rangle = (2\pi)^{-n} \langle \hat{u}\hat{v}, \hat{\phi} \rangle = (2\pi)^{-n} \langle \hat{u}, \hat{v}\hat{\phi} \rangle.$$

This holds for $\phi \in \mathscr{S}(\mathbf{R}^n)$, so in particular if $\phi \in C_c^\infty(\mathbf{R}^n)$. Then $\hat{v}\hat{\phi} = (v * \phi)\hat{}$, by Corollary 8.4.1. Hence it follows from (8.3.6) that

$$\langle w, \check{\phi} \rangle = \langle u, (v * \phi)\check{} \rangle = \langle u * v, \check{\phi} \rangle, \quad \phi \in C_c^\infty(\mathbf{R}^n).$$

Thus $w = u * v$ in $\mathscr{D}'(\mathbf{R}^n)$, and as $w \in \mathscr{S}'(\mathbf{R}^n)$, the theorem is implied by this.

Remark. It is evident that $\mathscr{O}_M(\mathbf{R}^n)$ is a subspace of $\mathscr{S}'(\mathbf{R}^n)$, as its members are continuous functions of polynomial growth. Hence

$$\mathscr{O}'_C(\mathbf{R}^n) = \mathscr{F}^{-1}\, \mathscr{O}_M(\mathbf{R}^n) \tag{8.4.4}$$

is also a subspace of $\mathscr{S}'(\mathbf{R}^n)$. If $u \in \mathscr{S}'(\mathbf{R}^n)$ and $v \in \mathscr{O}'_C(\mathbf{R}^n)$, then $\hat{u}\hat{v} \in \mathscr{S}'(\mathbf{R}^n)$. So one can *define* $u * v$ in this case as $\mathscr{F}^{-1}(\hat{u}\hat{v})$, and this agrees with the usual definition of $u * v$ when v has compact support.

One can also use Lemma 8.4.1 to prove another version of the structure theorem for $\mathscr{D}'(\mathbf{R}^n)$. One needs the following simple lemma:

Lemma 8.4.2. If $g \in L_1(\mathbf{R}^n)$ then

$$\mathscr{F}^{-1} g = (2\pi)^{-n} \int e^{ix\cdot\xi} g(\xi)\, d\xi, \tag{8.4.5}$$

which is a bounded continuous function.

Proof. Since $L_1(\mathbf{R}^n) \subset \mathscr{S}'(\mathbf{R}^n)$, the inverse Fourier transform $\mathscr{F}^{-1} g = f \in \mathscr{S}'(\mathbf{R}^n)$ exists, and by (8.3.6) one has

$$\langle f, \check{\phi} \rangle = (2\pi)^{-n} \langle \hat{f}, \hat{\phi} \rangle = (2\pi)^{-n} \int \hat{f}(\xi)\hat{\phi}(\xi)\, d\xi, \quad \phi \in \mathscr{S}(\mathbf{R}^n).$$

So, by Fubini's theorem,

$$\langle f, \check{\phi} \rangle = (2\pi)^{-n} \int \check{\phi}(x)\, dx \int e^{ix\cdot\xi} g(\xi)\, d\xi, \quad \phi \in \mathscr{S}(\mathbf{R}^n),$$

and this gives (8.4.5) when ϕ is replaced by $\check{\phi}$, in view of Theorem 8.1.3 (i).

Note that (8.4.5) is just the classical inversion formula (8.2.6), but that $\mathscr{F}^{-1} g$ need not be in $L_1(\mathbf{R}^n)$.

Theorem 8.4.3. If $u \in \mathscr{D}'(\mathbf{R}^n)$, and $X \subset \mathbf{R}^n$ is a bounded open set, then there is an $f \in C^0(\mathbf{R}^n)$ and an integer $N \geqslant 0$ such that $u = (1 - \Delta)^N f$ on X.

Proof. By the hypothesis, one can choose a cut-off function $\rho \in C_c(\mathbf{R}^n)$ such that $\rho = 1$ on X, and one then has $u = \rho u$ on X. As $\rho u \in \mathscr{E}'(\mathbf{R}^n)$, Lemma 8.4.1 implies that its Fourier transform is a continuous (indeed, C^∞) function of polynomial growth. Hence one can find a nonnegative integer N such that

$$g(\xi) = (1 + |\xi|^2)^{-N} (\rho u)\hat{}(\xi) \in L_1(\mathbf{R}^n).$$

By Lemma 8.4.2, one has $g=\hat{f}$, where f is a bounded continuous function on $\mathbf{R}^n$. Also, $((1-\Delta)^N f)\hat{} = (1+|\xi|^2)^N g$ by (8.3.7), so $(\rho u)\hat{} = ((1-\Delta)^N f)\hat{}$ whence $\rho u = (1-\Delta)^N f$, and so we are done.

8.5. Poisson's summation formula, and periodic distributions

A distribution $u \in \mathscr{D}'(\mathbf{R}^n)$ will be called *periodic* if

$$\tau_g u = u, \quad g \in \mathbf{Z}^n. \tag{8.5.1}$$

For this to hold, it is of course sufficient to have $\tau_{e_j} u = u$ for $j=1,\ldots,n$, where the e_j are the coordinate vectors $(1,0,\ldots,0),\ldots,(0,\ldots,0,1)$. It will be shown in this subsection that a periodic distribution can be expanded as a Fourier series. We first prove a technical lemma.

Lemma 8.5.1. Put $J=\{x: |x_1|<1,\ldots,|x_n|<1\}$. There is a $\psi \in C_c^\infty(\mathbf{R}^n)$ such that

$$\psi \geqslant 0, \quad \operatorname{supp} \psi \subset J, \quad \sum_{g \in \mathbf{Z}^n} \tau_g \psi = 1, \tag{8.5.2}$$

so that $(\tau_g \psi)_{g \in \mathbf{Z}^n}$ is a partition of unity subordinated to the covering $\{J+g : g \in \mathbf{Z}^n\}$.

Proof. Choose some $\delta \in (\frac{1}{2},1)$, and a cut-off function $\psi_0(t) \in C_c^\infty(\mathbf{R})$ such that $\psi_0 \geqslant 0$, $\psi_0 = 1$ for $|t|<\delta$, and $\operatorname{supp}\psi_0 \subset (-1,1)$. Evidently at least one, and at most two, of the functions $\psi_0(t-k)$, where $k \in \mathbf{Z}$, is then positive at each point $t \in \mathbf{R}$. From this it is easy to prove that

$$\psi_1(x) = \psi_0(x) \Big/ \sum_{k=-\infty}^{\infty} \psi_0(x-k)$$

satisfies (8.5.2) when $n=1$. Now set

$$\psi(x) = \psi_1(x_1)\ldots\psi_1(x_n), \quad x \in \mathbf{R}^n;$$

then trivial calculations show that (8.5.2) holds, and the lemma is proved.

We now consider the periodic distribution whose restriction to J is the Dirac distribution.

Theorem 8.5.1. (Poisson's summation formula.) The following identity holds in $\mathscr{S}'(\mathbf{R}^n)$:

$$\sum_{g \in \mathbf{Z}^n} \tau_g \delta = \sum_{g \in \mathbf{Z}^n} \exp(2\pi i g \cdot x). \tag{8.5.3}$$

Proof. It is obvious that the first member of (8.5.3) converges to a distribution which satisfies (8.5.1). Let us consider the second member; denote it by v.

Formally,

$$\langle v, \phi \rangle = \sum_{g \in \mathbf{Z}^n} \hat{\phi}(-2\pi g), \quad \phi \in \mathscr{S}(\mathbf{R}^n). \tag{8.5.4}$$

Now

$$\sum_{g \in \mathbf{Z}^n} (1 + |g|)^{-n-1} < \infty, \quad \int (1 + |x|)^{-n-1}\, dx < \infty.$$

So, if $\phi \in \mathscr{S}(\mathbf{R}^n)$, then it follows from (8.2.3) that for any $m \geqslant 0$ there is a $C_m \geqslant 0$ such that

$$|g|^m |\hat{\phi}(-2\pi g)| \leqslant C_m \sum_{|\alpha| = m} \sup |(1 + |x|)^{n+1} |D^\alpha \phi|$$

and it is then clear from (8.5.4) that

$$v = \sum_{g \in \mathbf{Z}^n} e^{2\pi i g \cdot x} \in \mathscr{S}'(\mathbf{R}^n). \tag{8.5.5}$$

It is an easy exercise to prove that translation is a continuous map $\mathscr{S}'(\mathbf{R}^n) \to \mathscr{S}'(\mathbf{R}^n)$; hence it follows from (8.5.5) that v is a periodic distribution. So one has, with ψ as in Lemma 8.5.1,

$$v = \sum_{g \in Z^n} v\tau_g \psi = \sum_{g \in Z^n} \tau_g(v\psi). \tag{8.5.6}$$

It also follows from (8.5.5) that

$$(e^{2\pi i g \cdot x} - 1)v = 0, \quad g \in \mathbf{Z}^n.$$

This carries over to $v\psi$, in particular one has

$$(e^{2\pi i x_j} - 1)(v\psi) = 0, \quad j = 1, \ldots, h.$$

Now, if $t \in \mathbf{R}$, then

$$e^{2\pi i t} - 1 = 2ie^{\pi i t} \sin \pi t = t\sigma(t)$$

say, where $\sigma(t) \in C^\infty(\mathbf{R})$ does not vanish on $(-1, 1)$. So

$$x_j v\psi = 0, \quad j = 1, \ldots, n. \tag{8.5.7}$$

If $\phi \in C^\infty(\mathbf{R}^n)$ then, by (3.2.6),

$$\phi(x) = \phi(0) + \sum_{j=1}^{n} x_j \phi_j(x),$$

where $\phi_j \in C^\infty(\mathbf{R}^n)$, $j = 1, \ldots, n$. As $v\psi$ has compact support it thus follows from (8.5.7) that

$$\langle v\psi, \phi \rangle = \langle v\psi, 1 \rangle \phi(0) = C\phi(0),$$

which means that $v\psi = C\delta$; hence (8.5.6) becomes

$$v = C \sum_{g \in \mathbf{Z}^n} \tau_g \delta. \tag{8.5.8}$$

So it only remains to show that $C = 1$. Let χ be the characteristic function of the unit cube $I = (0, 1)^n$; as $v \in \mathscr{S}'(\mathbf{R}^n)$ and $\chi \in \mathscr{E}'(\mathbf{R}^n)$, the convolution $v * \chi$ exists, and can be computed from (8.5.5),

$$v * \chi = \sum_{g \in \mathbf{Z}^n} \chi * e^{2\pi i g \cdot x} = 1$$

(the only nonzero term comes from $g = (0, \ldots, 0)$). On the other hand, (8.5.8) gives

$$v * \chi = C \sum_{g \in \mathbf{Z}^n} \tau_g \chi = C$$

neglecting a set of measure zero. Hence $C = 1$, and so (8.5.3) is proved.

Corollary 8.5.1. If $u \in \mathscr{E}'(\mathbf{R}^n)$, then

$$\sum_{g \in \mathbf{Z}^n} \tau_g u = \sum_{g \in \mathbf{Z}^n} \hat{u}(2\pi g) e^{2\pi i x \cdot g} \tag{8.5.9}$$

Proof. As $v \in \mathscr{S}'(\mathbf{R}^n)$, one can equate the convolution of u and the two members of (8.5.3).

Strictly speaking, it is the identity (8.5.9) rather than the special case (8.5.3) which is called Poisson's summation formula. It actually holds for a more extensive class of distributions, for example $\mathscr{O}'_C(\mathbf{R}^n)$. We shall now use it to obtain the Fourier series expansion of a periodic distribution.

Theorem 8.5.2. Let $u \in \mathscr{D}'(\mathbf{R}^n)$ be a periodic distribution, so that $\tau_g u = u$ for all $g \in \mathbf{Z}^n$. Then $u \in \mathscr{S}'(\mathbf{R}^n)$, and

$$u = \sum_{g \in \mathbf{Z}^n} \hat{u}_g e^{2\pi i g \cdot x} \tag{8.5.10}$$

where the $\hat{u}_g$ are complex numbers such that

$$|\hat{u}_g| \leqslant C(1 + |g|)^N, \quad g \in \mathbf{Z}^n, \tag{8.5.11}$$

for some $C, N \geqslant 0$.

Proof. Let ψ be as in Lemma 8.5.1. Then

$$u = \sum_{g \in \mathbf{Z}^n} u \tau_g \psi = \sum_{g \in \mathbf{Z}^n} \tau_g (u\psi).$$

Thus u is the convolution of $u \in \mathscr{E}'(\mathbf{R}^n)$ and of the first member of (8.5.3), which is a tempered distribution by (the proof of) Theorem 8.5.1; hence $u \in \mathscr{S}'(\mathbf{R}^n)$. Moreover, Poisson's summation formula gives

$$u = \sum_{g \in \mathbf{Z}^n} (u\psi)\hat{}(2\pi g) \mathrm{e}^{2\pi i g \cdot x},$$

which is (8.5.10) with

$$\hat{u}_g = (u\psi)\hat{}(2\pi g) = \langle u(x)\psi(x), \mathrm{e}^{-2\pi i g \cdot x} \rangle.$$

The inequalities (8.5.11) now follow from Lemma 8.4.1, and so the theorem is proved.

By (8.5.11), the series (8.5.10) converges in $\mathscr{S}'(\mathbf{R}^n)$, so

$$\hat{u} = \sum_{g \in \mathbf{Z}^n} \hat{u}_g \delta_{2\pi g}. \tag{8.5.12}$$

The $\hat{u}_g$ are therefore determined by u, irrespective of the choice of the cut-off function ψ. One can also derive a more explicit formula for them. Let χ be the characteristic function of the unit cube $I^n = (0, 1)^n$. Then, if $g, g' \in \mathbf{Z}^n$, one has

$$(\chi(x)\mathrm{e}^{2\pi i g \cdot x}) * \mathrm{e}^{2\pi i g' \cdot x} = \int_{I^n} \mathrm{e}^{2\pi i g' \cdot x} \mathrm{e}^{2\pi i (g - g') \cdot t} \, \mathrm{d}t$$

$$= \begin{cases} \mathrm{e}^{2\pi i g \cdot x} & \text{if } g' = g, \\ 0 & \text{if } g' \neq g. \end{cases}$$

Hence, as (8.5.10) converges in $\mathscr{S}'(\mathbf{R}^n)$ and so also in $\mathscr{D}'(\mathbf{R}^n)$, in Theorem 5.1.3 (i) implies that

$$u(x) * (\chi(x)\mathrm{e}^{2\pi i g \cdot x}) = \hat{u}_g \mathrm{e}^{2\pi i g \cdot x}, \quad g \in \mathbf{Z}^n. \tag{8.5.13}$$

Theorem 8.5.2 also has a converse:

Theorem 8.5.3. Let $(c_g)_{g \in \mathbf{Z}^n}$ be complex numbers which satisfy the inequalities

$$|c_g| \leqslant C(1 + |g|)^N, \quad g \in \mathbf{Z}^n \tag{8.5.14}$$

for some $C, N \geqslant 0$. Then the series

$$u = \sum_{g \in \mathbf{Z}^n} c_g \mathrm{e}^{2\pi i g \cdot x} \tag{8.5.15}$$

converges in $\mathscr{S}'(\mathbf{R}^n)$, and its sum satisfies (8.5.1), that is to say it is a periodic distribution.

Proof. Straightforward, and left as an exercise.

8.6. The elliptic regularity theorem

To conclude this chapter, the Fourier transform will now be used to prove an important result in the theory of linear partial differential equations. We need a preliminary definition, and a lemma.

Definition 8.6.1. Let $X \subset \mathbf{R}^n$ be an open set, and let $u \in \mathscr{E}'(X)$. The singular support of u is the complement in X of

$$\{x \in X: u \text{ is } C^\infty \text{ on some neighbourhood of } x\}.$$

It is denoted by sing supp u.

Note that u is C^∞ in the complement of its singular support, and that this is the largest open subset of X for which this is true; this is proved by the usual partition of unity argument. The singular support itself is a closed subset of X.

Lemma 8.6.1. Let $u \in \mathscr{D}'(\mathbf{R}^n)$ and $v \in \mathscr{E}'(\mathbf{R}^n)$. Then

$$\text{sing supp } u * v \subset \text{sing supp } u + \text{sing supp } v. \tag{8.6.1}$$

Proof. Choose functions $\rho \in C^\infty(\mathbf{R}^n)$ and $\psi \in C_c^\infty(\mathbf{R}^n)$ such that $\rho = 1$ on a neighbourhood of sing supp u and $\psi = 1$ on a neighbourhood of sing supp v. Then

$$\begin{aligned} u * v &= (\rho u + (1-\rho)u) * (\psi v + (1-\psi)v) \\ &= \rho u * \psi v + \rho u * (1-\psi)v + (1-\rho)u * v + (1-\rho)u * (1-\psi)v. \end{aligned}$$

Each of the convolutions other than $\rho u * \psi v$ on the right hand side has at least one C^∞ factor, so is C^∞. Hence

$$\text{sing supp } u * v \subset \text{supp } \rho u + \text{supp } \psi v \subset \text{supp } \rho + \text{supp } \psi,$$

and (8.6.1) now follows if one takes the intersection of the second members over all admissible functions ρ and ψ.

Let

$$P(D) = \sum_{|\alpha| \leq m} a_\alpha D^\alpha \tag{8.6.2}$$

be a linear differential operator of order m, with constant coefficients, defined on $\mathbf{R}^n$. The polynomial

$$P(\xi) = \sum_{|\alpha| \leq m} a_\alpha \xi^\alpha$$

is called the *symbol* of P, and the sum of the terms of order m in $P(\xi)$ is the *principal symbol* of P,

$$\sigma_P(\xi) = \sum_{|\alpha| \leq m} a_\alpha \xi^\alpha \tag{8.6.3}$$

Note that σ_P is a homogeneous polynomial of degree m.

By definition, the differential operator P is called *elliptic* if

$$\sigma_P(\xi) \neq 0 \quad \text{for } 0 \neq \xi \in \mathbf{R}^n. \tag{8.6.4}$$

The Laplacian is the archetypal elliptic operator; another, equally important, example is the Cauchy-Riemann operator $\frac{1}{2}(\partial_1 + i\partial_2)$ on $\mathbf{R}^2$.

Theorem 8.6.1. (The elliptic regularity theorem.) Let $X \subset \mathbf{R}^n$ be an open set, and let P be an elliptic operator with constant coefficients. Then

$$\text{sing supp } u = \text{sing supp } Pu \tag{8.6.5}$$

for all $u \in \mathscr{D}'(X)$.

Proof. It follows from the hypothesis (8.6.4) that $\sigma_P(\xi) \neq 0$ when $\xi \in \mathbf{S}^{n-1}$. As the unit sphere is compact, this implies

$$\inf \{|\sigma_P(\xi)|: \xi \in \mathbf{S}^{n-1}\} > 0$$

So the homogeneity of σ_P gives the inequality

$$|\sigma_P(\xi)| \geqslant c|\xi|^m, \quad \xi \in \mathbf{R}^n, \tag{8.6.6}$$

where $c > 0$. It follows that

$$|P(\xi)| \geqslant c|\xi|^m - c_1|\xi|^{m-1} - \ldots - c_m \tag{8.6.7}$$

where $c_1, \ldots, c_m$ are nonnegative real numbers. This in turn shows that, given any $\delta > 0$, one can find $t > 0$ such that

$$|P(\xi)| \geqslant \delta \quad \text{if } |\xi| \geqslant t. \tag{8.6.8}$$

Let $\chi \in C_c^\infty(\mathbf{R}^n)$ be such that $0 \leqslant \chi \leqslant 1$ and $\chi = 1$ for $|\xi| \leqslant t$. Then

$$(1 - \chi(\xi))/|P(\xi)| \leqslant \delta^{-1}$$

by (8.6.8), so that $(1-\chi)/P \in \mathscr{S}'(\mathbf{R}^n)$. Denote its inverse Fourier transform by E, thus

$$\hat{E} = (1 - \chi(\xi))/P(\xi). \tag{8.6.9}$$

We now claim that $E \in C^\infty(\mathbf{R}^n \setminus \{0\})$. Indeed, one has

$$(x^\beta D^\alpha E)\hat{} = (-1)^{|\beta|} D^\beta(\xi^\alpha \hat{E})$$

for any pair of multi-indices α and β. Now it is easy to deduce from this and (8.6.6)-(8.6.8) that

$$D^\beta(\xi^\alpha \hat{E}) = 0(|\xi|^{|\alpha| - |\beta| - m}) \quad \text{as } |\xi| \to \infty;$$

the details are left as an exercise. Taking $|\beta| = |\alpha| + n - m + 1$ one thus gets $(x^\beta D^\alpha E)\hat{} \in L_1(\mathbf{R}^n)$, whence $x^\beta D^\alpha E \subset C^0(\mathbf{R}^n)$ if $|\beta| = |\alpha| + n - m + 1$; if $x \neq 0$ there is at least one such β for which $x^\beta \neq 0$, hence $D^\alpha E \in C^0(\mathbf{R}^n \setminus 0)$ for all α, and this proves the claim. Hence

$$\text{sing supp } E = \{0\}. \tag{8.6.10}$$

As $E \in \mathscr{S}'(\mathbf{R}^n)$, one has $(P(D)E)\hat{} = P(\xi)\hat{E}$, so it follows from (8.6.9) that

$$DE = \delta - \rho, \quad \rho = \mathscr{F}^{-1}\chi \in C^\infty(\mathbf{R}^n). \tag{8.6.11}$$

Let X' be a relatively compact open subset of X. (This means that $\bar{X}' \subset X$). Choose $\psi \in C_c^\infty(X)$ such that $\psi = 1$ on X'. Then, by (8.6.11),

$$\psi u = \delta * \psi u = E * P\psi u + \rho * \psi u, \tag{8.6.12}$$

as one can extend ψu trivially to an element of $\mathscr{E}'(\mathbf{R}^n)$. But $\rho * \psi u \in C^\infty(\mathbf{R}^n)$ since $\rho \in C^\infty(\mathbf{R}^n)$, and

$$\text{sing supp } E * P\psi u \subset \text{sing supp } P\psi u$$

by (8.6.10) and Lemma 8.6.1, so (8.6.12) gives

$$\text{sing supp } \psi u \subset \text{sing supp } P\psi u.$$

The opposite inclusion is trivial, so finally

$$\text{sing supp } \psi u = \text{sing supp } P\psi u.$$

As X' can be any relatively compact subset of X, this proves the theorem.

A distribution E which satisfies (8.6.11) is called a C^∞ parametrix of P. The last part of the proof shows that, for the purpose of proving a regularity theorem, a parametrix is as good as a fundamental solution. This principle, applied to Schwartz kernels and backed by an appropriate construction, gives the elliptic regularity theorem for differential operators with variable coefficients. The principal symbol is then still defined by (8.6.3), but the a_α are members of $C^\infty(X)$, and (8.6.4) is then required to hold on $X \times (\mathbf{R}^n \setminus \{0\})$.

Exercises

8.1. Prove that translation is a continuous map $\mathscr{S}(\mathbf{R}^n) \to \mathscr{S}(\mathbf{R}^n)$.

8.2. Show that the Fourier transform of a tempered distribution on $\mathbf{R}^n$ that is homogeneous of degree λ is a homogeneous distribution of degree $(-\lambda - n)$.

8.3. Let $u \in \mathscr{S}'(\mathbf{R}^n)$ and $v \in \mathscr{S}'(\mathbf{R}^m)$. Show that $u \otimes v \in \mathscr{S}'(\mathbf{R}^{n+m})$, and that $(u \otimes v)\hat{} = \hat{u} \otimes \hat{v}$.

8.4. Use semi-norm estimates to prove: (*a*) if $u \in \mathscr{S}'(\mathbf{R}^n)$ and $v \in C_c^\infty(\mathbf{R}^n)$, then $u * v \in \mathscr{O}_M(\mathbf{R}^n)$; (*b*) if $u \in \mathscr{S}'(\mathbf{R}^n)$ and $v \in \mathscr{E}'(\mathbf{R}^n)$, then $u * v \in \mathscr{S}'(\mathbf{R}^n)$.

8.5. Show that, if $u \in \mathscr{O}'_C(\mathbf{R}^n)$ and $\rho \in C_c^\infty(\mathbf{R}^n)$, then $u * \rho \in \mathscr{S}(\mathbf{R}^n)$.

8.6. Show that, if $u \in \mathscr{S}'(\mathbf{R}^n)$, and $\psi \in \mathscr{S}(\mathbf{R}^n)$ is such that $\hat{\psi} \in C_c^\infty(\mathbf{R}^n)$, then $(u\psi)\hat{} = (2\pi)^{-n}\hat{u} * \hat{\psi}$.

8.7. Let $x_+^{\lambda-1}$ be the distribution on $\mathbf{R}$ which is equal to the locally integrable function $x \mapsto x_+^{\lambda-1}$ when $\operatorname{Re} \lambda > 0$, and is defined by analytic continuation in λ for all $\lambda \in \mathbf{C} \setminus 0, -1, \ldots$. Show that $x_+^{\lambda-1} \in \mathscr{S}'(\mathbf{R})$, and that its Fourier transform is $\Gamma(\lambda)\mathrm{e}^{-\frac{1}{2}\pi\lambda\mathrm{i}}(\xi - \mathrm{i}0)^{-\lambda}$.

Deduce that, if $\lambda \notin \mathbf{Z}$, then

$$(x_+^{\lambda-1})\hat{} = \Gamma(\lambda)\,(\mathrm{e}^{-\frac{1}{2}\pi\lambda\mathrm{i}}\xi_+^{-\lambda} + \mathrm{e}^{\frac{1}{2}\pi\lambda\mathrm{i}}\xi_-^{-\lambda}).$$

8.8. Show that, in $\mathscr{S}'(\mathbf{R}^2)$,

$$\mathscr{F}\left(\frac{1}{x_1 + ix_2}\right) = \frac{2\pi i}{\xi_1 + i\xi_2}.$$

8.9. Let c be a complex number, with $\operatorname{Im} c > 0$. Prove that

$$\mathscr{F}(\exp(\tfrac{1}{2}icx^2)) = \left(\frac{2\pi}{c}\right)^{\frac{1}{2}} \exp\left(i\frac{\pi}{4} - \frac{i\xi^2}{2c}\right),$$

where $c^{\frac{1}{2}}$ is chosen such that $\operatorname{Im} c^{\frac{1}{2}} > 0$. By considering the limit in $\mathscr{S}'(\mathbf{R})$ as $\operatorname{Im} c \to 0+$, deduce that, if $\omega \in \mathbf{R}$, then

$$\mathscr{F}(\exp(\tfrac{1}{2}i\omega x^2)) = \left(\frac{2\pi}{|\omega|}\right)^{\frac{1}{2}} \exp\left(i\frac{\pi}{4}\operatorname{sign}\omega - \frac{i\xi^2}{2\omega}\right).$$

Now let A be a real symmetric nonsingular $n \times n$ matrix. Show that, in $\mathscr{S}'(\mathbf{R}^n)$,

$$\mathscr{F}(\exp(\tfrac{1}{2}i(Ax \cdot x)) = \frac{(2\pi)^{\frac{1}{2}n}}{|\det A|^{\frac{1}{2}}} \exp\left(i\frac{\pi}{4}\operatorname{sgn} A - \tfrac{1}{2}i(A^{-1}\xi \cdot \xi)\right),$$

where $\operatorname{sgn} A$ (the signature of A) is $\Sigma_{j=1}^{n} \operatorname{sign} \lambda_j$, the λ_j being the eigenvalues of A.

8.10. Let t be a positive real variable, let $\omega_1, \ldots, \omega_n$ be real numbers, none of which are zero, and let $\phi \in \mathscr{S}(\mathbf{R}^n)$. Put

$$F(t) = \int \phi(x) \exp\left(\frac{1}{2}it \sum_{j=1}^{n} \omega_j x_j^2\right) dx.$$

Show that, for any $N \geqslant 1$,

$$F(t) = \frac{(2\pi)^{\frac{1}{2}n}}{|\omega_1 \ldots \omega_n|^{\frac{1}{2}}} \exp\left(\frac{i\pi}{4} \sum_{j=1}^{n} \operatorname{sign} \omega_j\right) t^{-\frac{1}{2}n}$$
$$\left(\sum_{k=0}^{N-1} \frac{1}{k!}\left(\frac{1}{2it}\right)^k (Q(D))^k \phi(x)|_{x=0} + r_N(x,t)\right),$$

where

$$Q(\xi) = \sum_{j=1}^{n} \omega_j^{-1} \xi_j^2,$$

and the remainder can be estimated by

$$|r_N| \leqslant C_N t^{-N} \sum_{|\alpha| \leqslant n+1} \|\Delta^N D^\alpha \phi\|_{L_1(\mathbf{R}^n)},$$

C_N being a constant, and Δ the Laplacian. (This is the basis of the 'method of stationary phase' for obtaining asymptotic expansions.)

8.11. Let A be a real nonsingular $n \times n$ matrix, and let

$$\Gamma = \{Ag : g \in \mathbf{Z}^n\}$$

be the lattice in $\mathbf{R}^n$ generated by the columns of A, regarded as vectors in $\mathbf{R}^n$. Prove that

$$\sum_{\gamma \in \Gamma} \tau_\gamma \delta = |\det A|^{-1} \sum_{\gamma' \in \Gamma'} \exp(2\pi i \gamma' \cdot x),$$

where Γ' is the lattice dual to Γ,

$$\Gamma' = \{\gamma' \in \mathbf{R}^n : \gamma \cdot \gamma' \in \mathbf{Z} \text{ for all } \gamma \in \Gamma\}.$$

8.12. Let $P(D)$ be a linear differential operator with constant coefficients on $\mathbf{R}^n$.

(i) Show that, if $u \in \mathcal{E}'(\mathbf{R}^n)$ and $Pu = 0$, then $u = 0$.

(ii) Show that, if $P(\xi) \neq 0$ when $0 \neq \xi \in \mathbf{R}^n$, then any tempered distribution satisfying $Pu = 0$ is a polynomial.

(iii) If $n = 2$ and $P(\xi) = \xi_1 + i\xi_2$, determine all $u \in \mathcal{S}'(\mathbf{R}^2)$ such that $Pu = 0$.

8.13. Let $k \in C^\infty(\mathbf{R}^n \setminus \{0\})$ be homogeneous of degree $-n$, and suppose that

$$\int_{S^{n-1}} k(\theta)\, d\omega(\theta) = 0.$$

Denote the principal value distribution determined by k (Exercise 2.5) by K:

$$\langle K, \phi \rangle = \lim_{\epsilon \to 0+} \int_{|x| > \epsilon} k(x)\phi(x)\, dx, \quad \phi \in C_c^\infty(\mathbf{R}^n).$$

Show that $K \in \mathcal{S}'(\mathbf{R}^n)$, and that $\hat{K} \in C^\infty(\mathbf{R}^n \setminus \{0\})$, and that

$$\int_{S^{n-1}} \hat{K}(\theta)\, d\omega(\theta) = 0.$$

8.14. This, and the following, exercise give an outline of a method for obtaining a fundamental solution of a homogeneous elliptic differential operator on $\mathbf{R}^n$.

Let $k \in C^\infty(\mathbf{R}^n \setminus \{0\})$ be homogeneous of degree $(-N-n)$, where N is a non-negative integer. Prove that one can define a distribution $K \in \mathcal{S}'(\mathbf{R}^n)$ as the finite part of a divergent integral,

$$\langle K, \phi \rangle = Pf \int k(x)\phi(x)\, dx$$

$$= \lim_{\epsilon \to 0+} \int_{|x| > \epsilon} k(x)\phi(x)\, dx + \sum_{j=0}^{N-1} c_j \epsilon^{j-N} + c_N \log \epsilon, \qquad \phi \in C_c^\infty(\mathbf{R}^n),$$

where $c_0, c_1, \ldots, c_N$ are constants, chosen so that the limit exists. Show also that, if $t > 0$, and Λ_t is the distribution obtained from K by a dilatation t, then

$$\langle K_t, \phi \rangle = t^{-N-h} \langle K, \phi \rangle + \frac{t^{-N-n}}{N!} \log t \int_{S^{n-1}} k(\theta) \left(\sum_{j=1}^{n} \theta_j \partial_j \right) \phi|_{x=0}\, d\omega(\theta).$$

Deduce that

$$\hat{K}(\xi) = \hat{H}(\xi) - Q(\xi) \log |\xi|,$$

where

$$Q(\xi) = \frac{1}{N!} \sum_{S^{n-1}} k(\theta)\, (-i\theta \cdot \xi)^N\, d\omega(\theta),$$

and $H \in \mathcal{S}'(\mathbf{R}^n)$ is homogeneous of degree N, and its restriction to $\mathbf{R}^n \setminus \{0\}$ is C^∞.

8.15. Let $P(\xi)$ be a homogeneous polynomial of degree m such that $P(\xi) \neq 0$ for $0 \neq \xi \in \mathbf{R}^n$. Show that the (elliptic) differential operator $P(D)$ has a fundamental solution E given by

$$\langle E, \phi \rangle = (2\pi)^{-n} Pf \int \frac{\hat{\phi}(\xi)}{P(\xi)}\, d\xi, \quad \phi \in C_c^\infty(\mathbf{R}^n).$$

Distinguish between the cases $m<n$ and $m \geqslant n$, and show that, if $m \geqslant n$, then E is the sum of a homogeneous function and of a term $Q(x) \log |x|$, where Q is a certain homogeneous polynomial.

8.16. Show that $u \in \mathscr{E}'(\mathbf{R}^n)$ is in $C^\infty(\mathbf{R}^n)$ (that is, $u \in C_c^\infty(\mathbf{R}^n)$) if and only if for each $N \geqslant 0$ there is a constant $C_N \geqslant 0$ such that

$$|\hat{u}(\xi)| \leqslant C_N(1+|\xi|)^{-N}. \qquad (*)$$

Call a subset Γ of $\mathbf{R}^n \setminus \{0\}$ conic if $\xi \in \Gamma$ implies that $t\xi \in \Gamma$ for all $t>0$. Define, for any $u \in \mathscr{E}'(\mathbf{R}^n)$, the closed conic set $W(u)$ to be the complement of

$$\{\xi \in \mathbf{R}^n \setminus \{0\}: \text{there is a conic neighbourhood of } \xi \text{ on which } (*) \text{ holds for some constants } (C_N)_{0 \leqslant N < \infty}.\}$$

Let $u \in \mathscr{E}'(\mathbf{R}^n)$ and $v \in \mathscr{E}'(\mathbf{R}^n)$ be such that $W(u)$ and $W(v)$ are disjoint subsets of $\mathbf{R}^n \setminus \{0\}$. Show that $u * v \in C_c^\infty(\mathbf{R}^n)$.

9 PLANCHEREL'S THEOREM, AND SOBOLEV SPACES

In the last section, the point of departure was the classical Fourier transform on $L_1(\mathbf{R}^n)$. One can extend the classical Fourier transform to the space $L_2(\mathbf{R}^n)$, which consists of (equivalence classes of) square integrable functions. Distribution theory gives a simple approach to this; it also allows one to introduce a family of spaces of functions or distributions, the L_2-based Sobolev spaces on $\mathbf{R}^n$, which plays an important part in the theory of partial differential equations. The setting for this theory is Hilbert space, and we begin with a summary of the relevant definitions and basic results.

9.1. Hilbert space

A *complex inner produce space* is a complex vector space H equipped with a positive definite Hermitian form (ϕ, ψ), called the *inner product* of ϕ and ψ. So (ϕ, ψ) is a function $H \times H \to \mathbf{C}$ with the following properties:

$$(\phi_1 + \phi_2, \psi) = (\phi_1, \psi) + (\phi_2, \psi), \quad (\lambda\phi, \psi) = \lambda(\phi, \psi) \quad \text{if } \lambda \in \mathbf{C},$$
$$(\psi, \phi) = \overline{(\phi, \psi)}, \quad (\phi, \phi) \geqslant 0, \quad (\phi, \phi) = 0 \quad \text{only if } \phi = 0.$$

Note that the inner product is linear in the first, and antilinear in the second, argument.

An important example is the space $L_2(X)$ consisting of the equivalence classes of square integrable functions on an open set $X \subset \mathbf{R}^n$. (This can be regarded as a space of distributions.) It becomes an inner product space if one sets

$$(\phi, \psi)_{L_2(X)} = \int_X \phi(x)\overline{\psi(x)}\, dx. \tag{9.1.1}$$

Given an inner product space H, one puts

$$\|\phi\| = (\phi, \phi)^{1/2}, \quad \phi \in H. \tag{9.1.2}$$

One then has an abstract Schwartz inequality:

$$|(\phi, \psi)| \leqslant \|\phi\|\, \|\psi\|, \quad \phi, \psi \in H. \tag{9.1.3}$$

This is trivial if $\phi = 0$ or if $(\phi, \psi) = 0$; if $\phi \neq 0$ it follows from

$$0 \leqslant \|c\phi - \psi\|^2 = |c|^2\, \|\phi\|^2 + \|\psi\|^2 - c(\phi, \psi) - \bar{c}(\psi, \phi)$$

with $c=(\psi,\phi)/\|\phi\|^2$. As a consequence of (9.1.3) one also has the triangle inequality

$$\|\phi+\psi\|\leqslant\|\phi\|+\|\psi\|,\quad \phi,\psi\in H. \tag{9.1.4}$$

As $\|\phi\|=0$ only when $\phi=0$, the map $\phi\mapsto\|\phi\|$ is thus a norm on H, and H is always given the topology induced by this norm.

A sequence $(\phi_j)_{1\leqslant j<\infty}$ is called convergent if there is a $\phi\in H$ such that $\|\phi-\phi_j\|\to 0$ as $j\to\infty$; it is called a Cauchy sequence if $\|\phi_j-\phi_k\|\to 0$ as $j,k=\infty$. By the triangle inequality (9.1.4), a convergent sequence is a Cauchy sequence. The converse statement need not be true; an example to this effect is the space $C_c^\infty(\mathbf{R}^n)$ equipped with the inner product (9.1.1). Now a normed space is a metric space (the metric is $\|\phi-\psi\|$); a metric space is called complete if every Cauchy sequence converges; a complete normed space is called a Banach space. A complete inner produce space is called a *Hilbert space*; note that it is also a Banach space.

The space $L_2(X)$, where $X\subset\mathbf{R}^n$ is an open set, and the inner product is given by (9.1.1), is a Hilbert space. This famous theorem will here be assumed without proof; see, for example, [5, p. 69].

A linear form u on H is continuous if and only if

$$|\langle u,\phi\rangle|\leqslant C\|\phi\|,\quad \phi\in H$$

for some constant C. The (strong) dual H^* of H is the vector space of continuous linear forms on H, equipped with the norm

$$\|u\|_{H^*}=\sup\{|\langle u,\phi\rangle|: \phi\in H,\ \|\phi\|_H=1\}. \tag{9.1.5}$$

(We use norms with subscripts when more than one space is involved.) For a given $\psi\in H$, the linear form $\phi\mapsto(\phi,\psi)$ is evidently a member of H^*; its norm is $\|\psi\|$, by (9.1.3) and (9.1.5). The *Riesz representation theorem* asserts that any continuous linear form on H can be realized in this way:

Theorem 9.1.1. Let H be a Hilbert space, and let u be a continuous linear form on H. Then there is a unique $\psi\in H$ such that

$$\langle u,\phi\rangle=(\phi,\psi),\quad \phi\in H, \tag{9.1.6}$$

and one has

$$\|\psi\|_H=\|u\|_{H^*}. \tag{9.1.7}$$

Proof. The theorem is trivial when $u=0$ (just take $\psi=0$), so we assume that $u\neq 0$. Then

$$\ker u=\{\phi\in H: \langle u,\phi\rangle=0\} \tag{9.1.8}$$

is a proper subspace of H, and by (9.1.3) it is a closed subspace. Hence, if we chose some $\phi_0\notin\ker u$, then

$$d=\inf\{\|\phi_0-\phi\|: \phi\in\ker u\} \tag{9.1.9}$$

is a positive real number. Let $(\phi_j)_{1 \leqslant j < \infty}$ be a sequence of elements of ker u such that $\|\phi_0 - \phi_j\| \to d$ as $j \to \infty$. Set $\phi_0 - \phi_j = \psi_j, j = 1, 2, \ldots$. By the easily verified 'parallelogram law', one has

$$\|\psi_j + \psi_k\|^2 + \|\psi_j - \psi_k\|^2 = 2(\|\psi_j\|^2 + \|\psi_k\|^2) \tag{9.1.10}$$

for all $j, k \geqslant 1$. The second member converges to $4d^2$ as $j, k \to \infty$. Also,

$$\|\psi_j + \psi_k\|^2 = 4\|\phi_0 - \tfrac{1}{2}(\phi_j + \phi_k)\|^2 \geqslant 4d^2,$$

as ϕ_j and ϕ_k are in ker u; so (9.1.10) implies that $\|\psi_j - \psi_k\| \to 0$ as $j, k \to \infty$. Since H is a Hilbert space, it follows that there is a $\psi_0 \in H$ such that $\psi_j \to \psi_0$. Clearly, $\|\psi_0\| = d$, whence $\psi_0 \notin$ ker u. Furthermore, by combining the two inequalities

$$d^2 = \|\psi_0\|^2 \leqslant \|\psi_0 \pm \phi\|^2, \quad \phi \in \ker u,$$

one readily sees that ψ_0 is orthogonal to ker u,

$$(\phi, \psi_0) = 0 \quad \text{if } \phi \in \ker u. \tag{9.1.11}$$

We can now prove the theorem. If $\phi \in H$, then

$$\phi\langle u, \psi_0\rangle - \psi_0\langle u, \phi\rangle \in \ker u,$$

so it follows from (9.1.11) that

$$\langle u, \psi_0\rangle (\phi, \psi_0) - \langle u, \phi\rangle \|\psi_0\|^2 = 0$$

which is (9.1.6), with $\psi = \psi_0\langle u, \psi_0\rangle / \|\psi_0\|^2$. Also, (9.1.3) now gives $\langle u, \phi\rangle \leqslant \|\phi\| \, \|\psi\|$, whence (9.1.7) if one takes $\phi = \psi$, and so we are done.

The Riesz representation theorem shows that a Hilbert space is isomorphic, and indeed isometric (as a Banach space) to its dual. The isometry is antilinear; to get a linear isometry, one can go over to the antidual of H, which is the vector space of continuous antilinear forms on H.

9.2. The Fourier transform on $L_2(\mathbf{R}^n)$

The members of $L_2(\mathbf{R}^n)$ are evidently tempered distributions, and as such they have Fourier transforms. It will be shown below that these are again in $L_2(\mathbf{R}^n)$. For the proof, one has to appeal to the fact that $C_c^\infty(\mathbf{R}^n)$ is dense in $L_2(\mathbf{R}^n)$. We thus have to establish this first, and so begin with a technical lemma.

Lemma 9.2.1. Let $\psi \in C_c^\infty(\mathbf{R}^n)$ be such that

$$\psi \geqslant 0, \quad \int \psi(x) \, dx = 1,$$

and set $\psi_\epsilon(x) = \epsilon^{-n} \psi(x/\epsilon)$, where $\epsilon > 0$. Let $f \in L_2(\mathbf{R}^n)$, and put

$$f_\epsilon(x) = f * \psi_\epsilon = \int f(y) \psi_\epsilon(x - y) \, dy. \tag{9.2.1}$$

Then $f_\epsilon \in C^\infty(\mathbf{R}^n)$, and $f_\epsilon \to f$ in $L_2(\mathbf{R}^n)$ as $\epsilon \to 0$. Also,

$$\|f_\epsilon\| \leqslant \|f\|, \quad \epsilon > 0. \tag{9.2.2}$$

Note. In (9.2.2), and throughout this subsection, $\|\cdot\|$ is the $L_2(\mathbf{R}^n)$ norm.

Proof. As

$$|f(y)\psi_\epsilon(x-y)| \leqslant \tfrac{1}{2}|f(y)|^2 + \tfrac{1}{2}|\psi_\epsilon(x-y)|^2,$$

the second member of (9.2.1) exists, and it follows from Theorem 5.2.1 that it is in $C^\infty(\mathbf{R}^n)$. By Schwartz's inequality,

$$|f_\epsilon(x)|^2 \leqslant \int |f(x-y)|^2 \psi_\epsilon(y)\,dy \int \psi_\epsilon(y)\,dy$$

$$= \int |f(x-y)|^2 \psi_\epsilon(y)\,dy.$$

Hence, by Fubini's theorem (Theorem 8.1.1)

$$\|f_\epsilon\|^2 \leqslant \int \psi_\epsilon(y)\,dy \int |f(x-y)|^2\,dx = \|f\|^2,$$

which is (9.2.2).

A similar manipulation gives

$$|f(x)-f_\epsilon(x)|^2 = \left|\int (f(x)-f(x-y))\psi_\epsilon(y)\,dy\right|^2$$

$$\leqslant \int |f(x)-f(x-y)|^2 \psi_\epsilon(y)\,dy,$$

whence, again by Fubini,

$$\|f-f_\epsilon\|^2 \leqslant \int \|\tau_y f - f\|^2 \psi_\epsilon(y)\,dy = \int \|\tau_{\epsilon y} f - f\|^2 \psi(y)\,dy, \tag{9.2.3}$$

making the change of variable $y \mapsto \epsilon y$. But translation is a continuous map $L_2(\mathbf{R}^n) \to L_2(\mathbf{R}^n)$. Indeed, it is well known that $C_c^0(\mathbf{R}^n)$ is dense in $L_2(\mathbf{R}^n)$ [5, p. 71]. So, given any $\delta > 0$, one can find $f_\delta \in C_c^\infty(\mathbf{R}^n)$ such that $\|f - f_\delta\| \leqslant \delta$. For any $h \in \mathbf{R}^n$ one then has

$$\|\tau_h f - f\| \leqslant 2\delta + \|\tau_h f_\delta - f_\delta\|.$$

Now $\|\tau_h f_\delta - f_\delta\| \to 0$ as $h \to 0$, by the uniform continuity of f_δ, and so it follows that $\|\tau_h f - f\| \to 0$ as $h \to 0$. We can apply this to (9.2.3), and observe also that

$$0 \leqslant \|\tau_{\epsilon y} f - f\|^2 \psi(y) \leqslant (\|f\| + \|\tau_{\epsilon y} f\|)^2 \sup \psi = 4\|f\|^2 \sup \psi.$$

So it follows by dominated convergence from (9.2.3) that $f_\epsilon \to f$ in $L_2(\mathbf{R}^n)$ as $\epsilon \to 0$, and the lemma is proved.

We can now prove the L_2-version of Theorem 1.2.1:

Theorem 9.2.1. $C_c^\infty(\mathbf{R}^n)$ is dense in $L_2(\mathbf{R}^n)$.

Proof. Let $f \in L_2(\mathbf{R}^n)$ be given, and define f_ϵ by (9.2.1).

Choose some $\phi \in C_c^\infty(\mathbf{R}^n)$ such that $0 \leqslant \phi \leqslant 1$ and $\phi = 1$ for $|x| \leqslant 1$, and put $g_\epsilon(x) = \phi(\epsilon x) f_\epsilon(x)$. Then $g_\epsilon \in C_c^\infty(\mathbf{R}^n)$, and

$$\begin{aligned}\|f - g_\epsilon\| &\leqslant \|(f - f_\epsilon)\phi(\epsilon x)\| + \|(\phi(\epsilon x) - 1)f\| \\ &\leqslant \|f - f_\epsilon\| + \left(\int_{|x| > \epsilon^{-1}} |f(x)|^2 \, dx\right)^{1/2}.\end{aligned}$$

In the last expression, the first term tends to 0 with ϵ by Lemma 9.2.1, and the second one does so because $f \in L_2(\mathbf{R}^n)$; so we are done.

We can now remark that

$$C_c^\infty(\mathbf{R}^n) \subset \mathscr{S}(\mathbf{R}^n) \subset L_2(\mathbf{R}^n) \subset \mathscr{S}'(\mathbf{R}^n), \tag{9.2.3}$$

where all the injections are continuous.

The next result is *Plancherel's theorem*; this is the key result in the theory.

Theorem 9.2.2. If $u \in L_2(\mathbf{R}^n)$, then its $\mathscr{S}'$-Fourier transform u is also a member of $L_2(\mathbf{R}^n)$, and their L_2-norms are related by Parseval's identity,

$$\|\hat{u}\|^2 = (2\pi)^n \|u\|^2. \tag{9.2.4}$$

Proof. We first establish (9.2.4) when $u \in \mathscr{S}(\mathbf{R}^n)$. If $\phi \in \mathscr{S}(\mathbf{R}^n)$ and $\psi \in \mathscr{S}(\mathbf{R}^n)$, then

$$\int \phi(x)\hat{\psi}(x) \, dx = \int \hat{\phi}(\xi)\psi(\xi) \, d\xi. \tag{9.2.5}$$

Take

$$\psi(\xi) = \overline{\hat{\phi}(\xi)} = \int \overline{\phi(x)} \, e^{ix\cdot\xi} \, dx.$$

Then it follows from the inversion formula that $\hat{\psi} = (2\pi)^n \bar{\phi}$, so (9.2.5) becomes

$$\|\hat{\phi}\|^2 = (2\pi)^n \|\phi\|^2, \quad \phi \in \mathscr{S}(\mathbf{R}^n). \tag{9.2.6}$$

If now $u \in L_2(\mathbf{R}^n)$, then its Fourier transform is the tempered distribution

$$\langle \hat{u}, \phi \rangle = \langle u, \hat{\phi} \rangle = \int u\hat{\phi} \, dx, \quad \phi \in \mathscr{S}(\mathbf{R}^n).$$

Hence, by Schwartz's inequality and (9.2.6),

$$|\langle \hat{u}, \phi \rangle| \leqslant \|u\| \, \|\hat{\phi}\| = (2\pi)^{\frac{1}{2}n} \|u\| \, \|\phi\|$$

By (9.2.3) and Theorem 9.2.1, this estimate extends to $\phi \in L_2(\mathbf{R}^n)$, so that $\phi \mapsto \langle \hat{u}, \phi \rangle$ extends to a continuous linear form on $L_2(\mathbf{R}^n)$ whose norm does not exceed $(2\pi)^{\frac{1}{2}n} \|u\|$. By the Riesz representation theorem (Theorem 9.1.1), there is thus a $v \in L_2(\mathbf{R}^n)$ such that

$$\langle \hat{u}, \phi \rangle = \int \phi \bar{v} \, dx, \quad \|v\| \leqslant (2\pi)^{\frac{1}{2}n} \|u\|, \phi \in L_2(\mathbf{R}^n).$$

So we have $\hat{u} = \bar{v} \in L_2(\mathbf{R}^n)$, and

$$\|\hat{u}\| \leqslant (2\pi)^{\frac{1}{2}n} \|u\|.$$

This in turn can be applied now to $\hat{u}$, and one obtains

$$\|\hat{\hat{u}}\| \leqslant (2\pi)^{\frac{1}{2}n} \|\hat{u}\| \leqslant (2\pi)^n \|u\|.$$

But $\hat{\hat{u}} = (2\pi)^n \check{u}$, and $\|\check{u}\| = \|u\|$. So we can conclude that $\|\hat{u}\| = (2\pi)^{\frac{1}{2}n} \|u\|$, and the theorem is proved.

Remark. If H is a Hilbert space, then an endomorphism which is both an isometry, and surjective (hence bijective), is called a *unitary operator* on H. In view of the Fourier inversion theorem on $\mathscr{S}'(\mathbf{R}^n)$, Theorem 9.2.2 thus implies that $(2\pi)^{-\frac{1}{2}} \mathscr{F}: L_2(\mathbf{R}^n) \to L_2(\mathbf{R}^n)$ is a unitary operator.

There is also a sesquilinear version of the Parseval identity (9.2.4) (which is also called Parseval's identity):

Corollary 9.2.1. If $u \in L_2(\mathbf{R}^n)$ and $v \in L_2(\mathbf{R}^n)$, then

$$(\hat{u}, \hat{v}) = (2\pi)^n (u, v). \tag{9.2.7}$$

Proof. This follows from (9.2.4) and the elementary identity

$$(u, v) = \tfrac{1}{4}(\|u + v\|^2 - \|u - v\|^2 + \mathrm{i}\|u + \mathrm{i}v\|^2 - \mathrm{i}\|u - \mathrm{i}v\|^2),$$

whose verification is left to the reader.

We shall now use this to derive the convolution theorem for $L_2(\mathbf{R}^n)$. We first prove

Lemma 9.2.2. Let $u \in L_2(\mathbf{R}^n)$ and $v \in L_2(\mathbf{R}^n)$. Then

$$u * v(x) = \int u(y) v(x - y) \, dy, \quad x \in \mathbf{R}^n, \tag{9.2.8}$$

is a bounded continuous function.

Proof. The integral exists, because

$$|u(y) v(x - y)| \leqslant \tfrac{1}{2} |u(y)|^2 + \tfrac{1}{2} |v(x - y)|^2.$$

Schwartz's inequality gives

$$|u * v(x)| \leqslant \|u\| \, \|v\| < \infty, \quad x \in \mathbf{R}^n,$$

and also, for any $h \in \mathbf{R}^n$,

$$|u * v(x + h) - u * v(x)| \leqslant \|u\| \, \|\tau_h v - v\|, \quad x \in \mathbf{R}^n.$$

It has already been shown, in the proof of Lemma 9.2.1, that translation is a continuous map $L_2(\mathbf{R}^n)\to L_2(\mathbf{R}^n)$; so the lemma follows.

The lemma shows that $u * v$, as a distribution, is in $\mathscr{S}'(\mathbf{R}^n)$, and so has a Fourier transform.

Theorem 9.2.3. If $u\in L_2(\mathbf{R}^n)$ and $v\in L_2(\mathbf{R}^n)$, then

$$(u * v)\hat{} = \hat{u}\hat{v}. \tag{9.2.9}$$

Proof. By Theorem 9.2.2, both $\hat{u}$ and $\hat{v}$ are members of $L_2(\mathbf{R}^n)$, and so it follows from

$$|\hat{u}\hat{v}| \leqslant \tfrac{1}{2}|\hat{u}|^2 + \tfrac{1}{2}|\hat{v}|^2$$

that $\hat{u}\hat{v}\in L_1(\mathbf{R}^n)$. Hence, by Lemma 8.4.2,

$$\mathscr{F}^{-1}(\hat{u}\hat{v})(x) = (2\pi)^{-n}\int \hat{u}(\xi)\hat{v}(\xi)e^{ix\cdot\xi}\,d\xi. \tag{9.2.10}$$

Take $z\in\mathbf{R}^n$ and observe that, as $v(\xi)\exp(iz\cdot\xi)\in L_2(\mathbf{R}^n)$, one can define $f_z\in L_2(\mathbf{R}^n)$ by

$$f_z = \mathscr{F}^{-1}(\overline{\hat{v}(\xi)}\,e^{-iz\cdot\xi}). \tag{9.2.11}$$

Then (9.2.10) and the Parseval identity (9.2.7) give

$$\mathscr{F}^{-1}(\hat{u}\hat{v})(z) = (u, f_z)_{L_2(\mathbf{R}^n)}. \tag{9.2.12}$$

It now remains to compute f_z. Now if $\phi\in\mathscr{S}(\mathbf{R}^n)$, then

$$\langle f_z, \check{\phi}\rangle = (2\pi)^{-n}\langle \hat{f}_z, \hat{\phi}\rangle = (2\pi)^{-n}\int \overline{\hat{v}(\xi)}\hat{\phi}(\xi)e^{-iz\cdot\xi}\,d\xi.$$

Another appeal to the Parseval identity (9.2.7) therefore gives

$$\langle f_z, \check{\phi}\rangle = \int \overline{v(x)}\phi(x-z)\,dx = \int \overline{v(z-x)}\check{\phi}(x)\,dx,$$

whence $f_z(x) = \overline{v(z-x)}$. So (9.2.12) becomes $\mathscr{F}^{-1}(\hat{u}\hat{v}) = u * v$, and we are done.

9.3. Sobolev spaces

Differentiability classes, and spaces of distributions of finite order, can be used to characterize the 'regularity' of a distribution. A different approach can be based on Theorem 9.2.2. It is more restricted, inasmuch as it does not apply to all of $\mathscr{D}'(\mathbf{R}^n)$. But it has proved to be of particular importance in the theory of partial differential equations. We shall outline this now, and begin with a definition.

Definition 9.3.1. Let $s\in\mathbf{R}$. The Sobolev space $H_s(\mathbf{R}^n)$ consists of all $u\in\mathscr{S}'(\mathbf{R}^n)$ such that $(1+|\xi|^2)^{\frac{1}{2}s}\hat{u}(\xi)\in L_2(\mathbf{R}^n)$.

Note that $\hat{u}$ is a function when $u \in H_s(\mathbf{R}^n)$, and that $H_0(\mathbf{R}^n) = L_2(\mathbf{R}^n)$.

Theorem 9.3.1. $H_s(\mathbf{R}^n)$ becomes a Hilbert space when it is equipped with the inner product

$$(u, v)_s = (2\pi)^{-n} \int (1 + |\xi|^2)^s \hat{u}(\xi)\overline{\hat{v}(\xi)}\, d\xi. \tag{9.3.1}$$

Furthermore, $\mathscr{S}(\mathbf{R}^n)$ is dense in $H_s(\mathbf{R}^n)$.

Proof. It is obvious that the integral on the right hand side of (9.3.1) exists, and is a positive semi-definite Hermitian form on H_s. That it is in fact positive definite, so that

$$\|u\|_s = (2\pi)^{-\frac{1}{2}n} \left(\int (1 + |\xi|^2)^{\frac{1}{2}s} |\hat{u}(\xi)|^2\, d\xi \right)^{1/2} \tag{9.3.2}$$

is a norm on $H_s(\mathbf{R}^n)$, follows from Theorem 9.2.2. The completeness of $H_s(\mathbf{R}^n)$ with respect to this norm is an immediate consequence of the completeness of $L_2(\mathbf{R}^n)$; the easy proof is left to the reader.

It is also clear that $\mathscr{S}(\mathbf{R}^n) \subset H_s(\mathbf{R}^n)$ for all $s \in \mathbf{R}$. To show that it is dense, let $u \in H_s(\mathbf{R}^n)$ be given. By Theorem 9.2.1, one can find a sequence $(\phi_j)_{1 \leqslant j < \infty} \in C_c^\infty(\mathbf{R}^n)$ such that $\phi_j \to (1 + |\xi|^2)^{\frac{1}{2}}\hat{u}$ in $L_2(\mathbf{R}^n)$ as $j \to \infty$. Then

$$\psi_j = \mathscr{F}^{-1}((1 + |\xi|^2)^{-\frac{1}{2}s}\phi_j), \quad j = 1, 2, \ldots$$

is in $\mathscr{S}(\mathbf{R}^n)$, as it is the inverse Fourier transform of a member of $C_c^\infty(\mathbf{R}^n)$. Also

$$\|u - \psi_j\|_s = (2\pi)^{-\frac{1}{2}n} \|(1 + |\xi|^2)^{\frac{1}{2}}\hat{u}(\xi) - \phi_j(\xi)\|_{L_2(\mathbf{R}^n)} \to 0$$

as $j \to \infty$, and so we are done.

Two immediate consequences of this theorem are worth recording.

Corollary 9.3.1. $H_s(\mathbf{R}^n) \subset H_t(\mathbf{R}^n)$ if $s \geqslant t$, and the inclusion map is continuous.

Corollary 9.3.2. If P is a linear differential operator with constant coefficients of order m, and $u \in H_s(\mathbf{R}^n)$, then $Pu \in H_{s-m}(\mathbf{R}^n)$, and the map $P: H_s \to H_{s-m}$ is continuous.

One can construct a pairing of $H_{-s}(\mathbf{R}^n)$ and $H_s(\mathbf{R}^n)$, as follows. Let $\phi \in \mathscr{S}(\mathbf{R}^n)$ and $\psi \in \mathscr{S}(\mathbf{R}^n)$. With ψ as the distribution, and ϕ as the test function, one has

$$\langle \psi, \phi \rangle = \int \psi(x)\phi(x)\, dx = (2\pi)^{-n} \int \hat{\psi}(\xi)\hat{\phi}(-\xi)\, d\xi,$$

using the Parseval identity (9.2.7). Now

$$\hat{\psi}(\xi)\hat{\phi}(-\xi) = (1 + |\xi|^2)^{-\frac{1}{2}s}\hat{\psi}(\xi)\, (1 + |\xi|^2)^{\frac{1}{2}s}\hat{\phi}(-\xi),$$

so it follows from Schwartz's inequality that

$$|\langle\psi, \phi\rangle| \leqslant \|\psi\|_{-s}\|\phi\|_s.$$

By Theorem 9.3.1, one can therefore extend $\langle\psi, \phi\rangle$ to a bilinear form on $H_{-s}(\mathbf{R}^n)\times H_s(\mathbf{R}^n)$,

$$\langle u, v\rangle = (2\pi)^{-n}\int \hat{u}(\xi)\hat{v}(-\xi)\,d\xi, \quad u\in H_{-s}(\mathbf{R}^n), \quad v\in H_s(\mathbf{R}^n) \tag{9.3.3}$$

which is continuous, since

$$|\langle u, v\rangle| \leqslant \|u\|_{-s}\|v\|_s. \tag{9.3.4}$$

Theorem 9.3.2. The pairing (9.3.3) gives a canonical isometric isomorphism of $H_{-s}(\mathbf{R}^n)$ and $(H_s(\mathbf{R}^n))^*$, the dual of $H_s(\mathbf{R}^n)$.

Proof. It follows from (9.3.4) that, for fixed $u\in H_{-s}(\mathbf{R}^n)$, $v\mapsto\langle u, v\rangle$ is a continuous linear form on $H_s(\mathbf{R}^n)$, whose norm does not exceed $\|u\|_{-s}$. Taking

$$v = v_0 = \mathscr{F}^{-1}((1+|\xi|^2)^{-s}\hat{u})\in H_s(\mathbf{R}^n),$$

one gets $\langle u, v_0\rangle = \|u\|_{-s}$. Hence the norm of $v\mapsto\langle u, v\rangle$ is equal to $\|u\|_{-s}$, and we have thus an isometry $H_{-s}(\mathbf{R}^n)\to(H_s(\mathbf{R}^n))^*$.

To prove that this isometry is surjective, and hence an isomorphism, let $u^*\in(H_s(\mathbf{R}^n))^*$. Then, by the Riesz representation theorem and (9.3.1) there is a $w\in H_s(\mathbf{R}^n)$ such that

$$u^*(v) = (v, w)_s = (2\pi)^{-n}\int(1+|\xi|^2)^s\hat{v}(\xi)\overline{\hat{w}(\xi)}\,d\xi.$$

But, clearly, if we set

$$u = \mathscr{F}^{-1}(\overline{\hat{w}(-\xi)}\,(1+|\xi|^2)^s)$$

then $u\in H_{-s}(\mathbf{R}^n)$ and $u^*(v) = \langle u, v\rangle$ for all $v\in H_s(\mathbf{R}^n)$, and the proof is complete.

When s is a positive integer, there are two other characterizations of $H_s(\mathbf{R}^n)$.

Theorem 9.3.3. Let m be a positive integer. Then

$$H_m(\mathbf{R}^n) = \{u\in\mathscr{D}'(\mathbf{R}^n)\colon D^\alpha u\in L_2(\mathbf{R}^n) \quad \text{if } |\alpha|\leqslant m\}. \tag{9.3.5}$$

Furthermore, $H_m(\mathbf{R}^n)$ is the completion of $C_c^\infty(\mathbf{R}^n)$ with respect to the norm

$$\|\phi\|^m = \left(\int\sum_{|\alpha|\leqslant m}|D^\alpha\phi|^2\,dx\right)^{1/2}. \tag{9.3.6}$$

Note. The second statement means that a sequence of functions of class $C_c^\infty(\mathbf{R}^n)$ which is a Cauchy sequence with respect to the norm (9.3.6) converges in $H_m(\mathbf{R}^n)$, and that every member of $H_m(\mathbf{R}^n)$ is the limit, in $H_m(\mathbf{R}^n)$, of such a sequence.

Proof. For $\xi \in \mathbf{R}^n$ one has

$$|\xi|^{2k} = \left(\sum_{j=1}^{n} \xi_j^2\right)^k = \sum_{|\alpha|=k} \frac{k!}{\alpha!} |\xi^\alpha|^2.$$

It is also obvious that $|\xi^\alpha|^2 \leqslant |\xi|^{2k}$ when $|\alpha| = k$. So there is a constant C_m such that

$$\sum_{|\alpha|\leqslant m} |\xi^\alpha|^2 \leqslant (1+|\xi|^2)^m \leqslant C_m \sum_{|\alpha|\leqslant m} |\xi^\alpha|^2. \tag{9.3.7}$$

If $u \in H_m(\mathbf{R}^n)$ then $u \in \mathscr{S}'(\mathbf{R}^n)$ and $\xi^\alpha \hat{u} = (D^\alpha u)\hat{} \in L_2(\mathbf{R}^n)$, by Definition 9.3.1 and (9.3.7); so u is in the subspace of $\mathscr{D}'(\mathbf{R}^n)$ defined by (9.3.5). On the other hand, any u which satisfies (9.3.5) is in $\mathscr{S}'(\mathbf{R}^n)$ (as it is in $L_2(\mathbf{R}^n)$), and such that $\xi^\alpha \hat{u} \in L_2(\mathbf{R}^n)$ for $|\alpha| \leqslant m$; so it follows from (9.3.7) that $u \in H_m(\mathbf{R}^n)$. This proves the first assertion of the theorem.

The inequalities (9.3.7) also show that the norms (9.3.2) and (9.3.6) are equivalent when $H_m(\mathbf{R}^n)$ is considered as a Banach space.

Turning to the second assertion, we consider a sequence $(\phi_j)_{1\leqslant j<\infty} \in C_c^\infty(\mathbf{R}^n)$ which is a Cauchy sequence with respect to the norm (9.3.6). Each sequence $(D^\alpha \phi_j)_{1\leqslant j<\infty}$, where $|\alpha| \leqslant m$, is Cauchy in $L_2(\mathbf{R}^n)$, so converges to an element u_α of $L_2(\mathbf{R}^n)$. By Schwartz's inequality, one has

$$\int \psi D^\alpha \phi_j \, dx = (-1)^{|\alpha|} \int \phi_j D^\alpha \psi \, dx \to (-1)^{|\alpha|} \int u_\alpha D^\alpha \psi \, dx$$

for all $\psi \in C_c^\infty(\mathbf{R}^n)$; hence $u_\alpha = D^\alpha u$ for $|\alpha| \leqslant m$, where u is the L_2-limit of the ϕ_j. So it follows from the first part of the theorem, which has already been proved, that $u \in H_m(\mathbf{R}^n)$.

To show that any $u \in H_m(\mathbf{R}^n)$ is the limit of such a Cauchy sequence of test functions, we form the regularizations u_ϵ of u, as in Lemma 9.2.1. We have already shown that $D^\alpha u$ is in $L_2(\mathbf{R}^n)$ if $|\alpha| \leqslant m$. Also,

$$\begin{aligned} D^\alpha u_\epsilon(x) &= \int u(y) D_x^\alpha \psi_\epsilon(x-y) \, dy = (-1)^{|\alpha|} \int u(y) D_y^\alpha \psi_\epsilon(x-y) \, dy \\ &= \int \psi_\epsilon(x-y) D^\alpha u(y) \, dy = D^\alpha u * \psi_\epsilon(x). \end{aligned}$$

Hence it follows from Lemma 9.2.1 that one has, in $L_2(\mathbf{R}^n)$,

$$\|D^\alpha u_\epsilon\| \leqslant \|D^\alpha u\|, \quad \|D^\alpha u_\epsilon - D^\alpha u\| \to 0 \quad \text{as } \epsilon \to 0. \tag{9.3.8}$$

Take $\phi \in C_c^\infty(\mathbf{R}^n)$ such that $0 \leqslant \phi \leqslant 1$ and $\phi = 1$ for $|x| \leqslant 1$, as in the proof of Theorem 9.2.1. Then, with $g_\epsilon(x) = \phi(\epsilon x) u_\epsilon(x)$,

$$D^\alpha(g_\epsilon - u) = \phi(\epsilon x)\,(D^\alpha u_\epsilon - D^\alpha u) + (\phi(\epsilon x) - 1)D^\alpha u$$
$$+ \sum_{0<\beta\leqslant\alpha} \frac{\alpha!}{\beta!(\alpha-\beta)!}\,\epsilon^{|\beta|}\phi^{(\beta)}(\epsilon x) D^{\alpha-\beta}u_\epsilon,$$

where $\phi^{(\beta)}(x) = D^\beta\phi(x)$. From this and (9.3.8) it follows that $D^\alpha g_\epsilon \to D^\alpha u$ in $L_2(\mathbf{R}^n)$ as $\epsilon\to 0$, for all α such that $|\alpha|\leqslant m$. But then $\|g_\epsilon - u\|^m \to 0$ as $\epsilon \to 0$, and the proof is complete.

Corollary 9.3.3. If m is a positive integer, then the members of $H_{-m}(\mathbf{R}^n)$ are finite sums of derivatives of order less than or equal to m of functions of class $L_2(\mathbf{R}^n)$.

Proof. By Theorem 9.3.1, every $u\in H_{-m}(\mathbf{R}^n)$ gives rise to a continuous linear form u^* on $H_m(\mathbf{R}^n)$. It follows from (9.3.7) that u^* is also a continuous linear form when $H_m(\mathbf{R}^n)$ is made into a Hilbert space by means of the inner product corresponding to the norm (9.3.6). The assertion now follows from the Riesz representation theorem; the details are left as an exercise.

The next theorem relates Sobolev spaces to differentiability classes; it is a simple version of the Sobolev embedding theorem. For this, we introduce the space $C_0^0(\mathbf{R}^n)$ of bounded continuous functions which tend to 0 as $|x|\to\infty$.

Theorem 9.3.4. If $s > \frac{1}{2}n$ and $u \in H_s(\mathbf{R}^n)$, then $u\in C_0^0(\mathbf{R}^n)$.

Note. Strictly speaking, one should say that u can be modified on a set of measure 0 so as to yield a function of class $C_0^0(\mathbf{R}^n)$.

Proof. If $s>\frac{1}{2}n$, then $\xi\mapsto(1+|\xi|^2)^{-s}$ is in $L_1(\mathbf{R}^n)$. Put

$$f(\xi) = (1+|\xi|^2)^{\frac{1}{2}s}\hat{u}(\xi).$$

Then $f\in L_2(\mathbf{R}^n)$, in fact $\|f\|_{L_2(\mathbf{R}^n)} = (2\pi)^{\frac{1}{2}n}\|u\|_s$. So, by Schwartz's inequality,

$$\|\hat{u}\|_{L_1(\mathbf{R}^n)} \leqslant \|f\|_{L_2(\mathbf{R}^n)}\left(\int (1+|\xi|^2)^{-s}\,d\xi\right)^{1/2},$$

whence

$$\|\hat{u}\|_{L_1(\mathbf{R}^n)} \leqslant C'\|u\|_s,$$

where C' is a constant. It follows therefore from Lemma 8.4.2 that u is the classical inverse Fourier transform of $\hat{u}$. Hence it is continuous and bounded, in fact

$$|u(x)| \leqslant C\|u\|_s$$

for some $C\geqslant 0$, and tends to zero as $|x|\to\infty$, by virtue of the Riemann-Lebesgue lemma.

Remark. The Riemann-Lebesgue lemma actually says that the Fourier transform of any $f\in L_1(\mathbf{R}^n)$ tends to zero as $|x|\to\infty$. This can be verified by computation when f is the characteristic function of a rectangle, and then follows in general because finite linear combinations of such functions are dense in $L_1(\mathbf{R}^n)$.

Corollary 9.3.4. If $s > \frac{1}{2}n + k$, where k is a positive integer, and $u \in H_s(\mathbf{R}^n)$, then u is equal, as a distribution, to a member of $C_c^k(\mathbf{R}^n)$, the space of functions f of class $C^k(\mathbf{R}^n)$ such that $D^\alpha f \to 0$ as $|x| \to \infty$ for $|\alpha| \leqslant k$.

Proof. Apply Theorem 9.3.4 to the (distribution) derivatives $D^\alpha u$, where $|\alpha| \leqslant k$, and then appeal to Theorem 2.1.2.

Finally, we shall prove that $H_s(\mathbf{R}^n)$ is stable under multiplication by rapidly decreasing test functions. We need two inequalities for this; the first one is a simple case of the Hausdorff-Young inequality:

Lemma 9.3.1. If $f \in L_2(\mathbf{R}^n)$ and $g \in L_1(\mathbf{R}^n)$, then

$$\|f * g\|_{L_2(\mathbf{R}^n)} \leqslant \|g\|_{L_1(\mathbf{R}^n)} \|f\|_{L_2(\mathbf{R}^n)}. \tag{9.3.9}$$

Proof. Suppose first that $f \in C_c^0(\mathbf{R}^n)$ and $g \in C_c^0(\mathbf{R}^n)$. Then

$$|f * g(x)|^2 \leqslant \int |g(y)|\,\mathrm{d}y \int |f(x-y)|^2 |g(y)|\,\mathrm{d}y$$

by Schwartz's inequality, so

$$\|f * g\|^2_{L_2(\mathbf{R}^n)} \leqslant \|g\|^2_{L_1(\mathbf{R}^n)} \|f\|^2_{L_2(\mathbf{R}^n)}$$

by the most elementary version of Fubini's theorem. As $C_c^0(\mathbf{R}^n)$ is dense in both $L_1(\mathbf{R}^n)$ and $L_2(\mathbf{R}^n)$, the inequality (9.3.9) now follows by continuity.

The second inequality is

$$(1 + |\xi|^2)^{\frac{1}{2}s} \leq (2(1 + |\eta|^2))^{\frac{1}{2}|s|}(1 + |\xi - \eta|^2)^{\frac{1}{2}s}, \xi \in \mathbf{R}^n, \eta \in \mathbf{R}^n \tag{9.3.10}$$

which follows from

$$(2(1 + |\eta|^2))^{-1}(1 + |\xi - \eta|^2) < 1 + |\xi|^2 < (1 + |\eta|^2)(1 + |\xi - \eta|^2).$$

To prove this, one observes first that the lower bound for $1 + |\xi|^2$ results from the estimate

$$1 + |\xi - \eta|^2 \leq 1 + 2(|\xi|^2 + |\eta|^2) < 2(1 + |\eta|^2)(1 + |\xi|^2).$$

The upper bound is then obtained by replacing ξ by $\eta - \xi$.

Theorem 9.3.5. If $u \in H_s(\mathbf{R}^n)$ and $\phi \in \mathscr{S}(\mathbf{R}^n)$, then $\phi u \in H_s(\mathbf{R}^n)$, and the map $u \mapsto \phi u$ is a continuous map $H_s(\mathbf{R}^n) \to H_s(\mathbf{R}^n)$, in fact

$$\|\phi u\|_s \leqslant (2\pi)^{-n} \|u\|_s \int (2(1 + |\eta|^2)^{\frac{1}{2}|s|} \hat{\phi}(\eta)|\,\mathrm{d}\eta.$$

Proof. All the assertions follow from (9.3.10). Suppose first that both ϕ and u are members of $\mathscr{S}(\mathbf{R}^n)$; then

$$(\phi u)\hat{} = (2\pi)^{-n}\hat{\phi} * \hat{u}.$$

Hence, by (9.3.10),

$$(1+|\xi|^2)^{\frac{1}{2}s}|(\phi u)\hat{}|$$
$$\leq (2\pi)^{-n}\int (2(1+|\eta|^2))^{\frac{1}{2}|s|}|\hat{\phi}(\eta)|(1+|\xi-\eta|^2)^{\frac{1}{2}s}|u(\xi-\eta)|\,d\eta,$$

so that (9.3.11) follows from (9.3.9). By Theorem 9.3.1, $\mathscr{S}(\mathbf{R}^n)$ is dense in $H_s(\mathbf{R}^n)$. The estimate (9.3.11) therefore extends by continuity to $u \in H_s(\mathbf{R}^n)$, and we are done.

Exercises

9.1. Let $1/x \in \mathscr{D}'(\mathbf{R})$ be the principal value distribution, and put $h = 1/\pi x$. By writing h as the sum of a distribution with compact support, and of a member of $L_2(\mathbf{R})$, show that the convolution $h * h$ is well defined, and compute it.

The Hilbert transform on $C_c^\infty(\mathbf{R})$ is the map $\phi \mapsto h * \phi$. Show that it extends to a unitary operator on $L_2(\mathbf{R})$.

9.2. (i) Show that, if $u \in \mathscr{E}'(\mathbf{R}^n)$ is of order m, then it is in $H_s(\mathbf{R}^n)$ if $s < -\frac{1}{2}n - m$.

(ii) Show that, in $\mathbf{R}^n$, the Dirac distribution δ belongs to $H_s(\mathbf{R}^n)$ if and only if $s < -\frac{1}{2}n$.

(iii) Show that if $u \in \mathscr{D}'(\mathbf{R}^n)$, if $u \not\equiv 0$, and if the support of u is a finite set of points, then u cannot be in $H_{-\frac{1}{2}n}(\mathbf{R}^n)$.

9.3. Let $k \in \mathscr{S}'(\mathbf{R}^n)$ be such that its Fourier transform is equal to a (measurable) function μ for which

$$\sup |(1+|\xi|^2)^{\frac{1}{2}t}\mu(\xi)| = M < \infty$$

for some real number t. Show that, if $s \in \mathbf{R}$ and $u \in H_s(\mathbf{R}^n)$, then $k * u$ is a well defined member of $H_{s+t}(\mathbf{R}^n)$, and that

$$\|k * u\|_{s+t} \leq M\|u\|_s.$$

9.4. Let $k \in C^1(\mathbf{R}^n \backslash \{0\})$ be homogeneous of degree $-n$, and satisfy

$$\int_{S^{n-1}} k(\theta)\,d\omega(\theta) = 0.$$

Let K be the principal value distribution corresponding to k (exercise 8.13). Show that, for any $s \in \mathbf{R}$, the convolution $K * u$ exists if $u \in H_s(\mathbf{R}^n)$, and that $u \mapsto K * u$ is a continuous map $H_s(\mathbf{R}^n) \to H_s(\mathbf{R}^n)$.

9.5. Let $(x, t) \in \mathbf{R}^n \times \mathbf{R}$ and $s \in \mathbf{R}$, and let $f(x) \in H_{s+1}(\mathbf{R}^n)$, $g(x) \in H_s(\mathbf{R}^n)$ be given. Put

$$v(\xi, t) = \hat{f}(\xi)\cos|\xi|t + \hat{g}(\xi)\frac{\sin|\xi|t}{|\xi|}, \quad (\xi, t) \in \mathbf{R}^n \times \mathbf{R},$$

where $\sin|\xi|t/|\xi|$ is defined by continuity for $\xi = 0$. Show that, for each $t \in \mathbf{R}$,

$\xi \mapsto v(\xi, t)$ is the Fourier transform of a distribution u_t such that $t \mapsto u_t$ is a C^∞ function $\mathbf{R} \to \mathscr{S}'(\mathbf{R}^n)$ and

$$(\partial_t^2 - \Delta)u_t = 0 \quad \text{if } t \in \mathbf{R}, \quad u_t = f, \quad \partial_t u_t = g \quad \text{if } t = 0.$$

Show also that

$$D_j u_t \in H_s(\mathbf{R}^n), \quad j = 1, \ldots, n, \quad D_t u_t \in H_s(\mathbf{R}^n),$$

and that

$$\sum_{j=1}^{n} \|D_j u\|_s^2 + \|D_t u\|_s^2 = \sum_{j=1}^{n} \|D_j f\|_s^2 + \|g\|_s^2$$

for all $t \in \mathbf{R}$.

9.6. Let $P(x, D)$ be a differential operator of order m on $\mathbf{R}^n$, with C^∞ coefficients. Let $K \subset \mathbf{R}^n$ be a compact set, and set, for any $s \in \mathbf{R}$,

$$H_s^K(\mathbf{R}^n) = \{u \in H_s(\mathbf{R}^n); \operatorname{supp} u \subset K\}.$$

Show that P is a continuous map $H_s^K(\mathbf{R}^n) \to H_{s-m}^K(\mathbf{R}^n)$.

9.7. Let $P(D)$ be an elliptic linear differential operator on $\mathbf{R}^n$, with constant coefficients. Show that there are positive real numbers C_0 and C_1 such that, if $u \in H_{s+m}(\mathbf{R}^n)$ and $Pu \in H_s(\mathbf{R}^n)$ for some $s \in \mathbf{R}$, and $t > 0$, then

$$\|u\|_{s+m} \leqslant C_0 \|Pu\|_s + C_1 \|u\|_{s+m-t}. \tag{$*$}$$

Deduce that, if (i) $u \in H_{s'}(\mathbf{R}^n)$ for some $s' \in \mathbf{R}$, and (ii) there is an $s \in \mathbf{R}$ such that $Pu \in H_s(\mathbf{R}^n)$, then $u \in H_{s+m}(\mathbf{R}^n)$, and $(*)$ holds.

Can one drop condition (i)?

10 THE FOURIER-LAPLACE TRANSFORM

The Laplace transform of a function f on $\mathbf{R}$ is

$$\int e^{-px} f(x)\, dx,$$

where p is a complex variable. If one puts $p = i\zeta$, it formally reduces to the Fourier transform of f,

$$\int e^{-i\zeta x} f(x)\, dx, \tag{*}$$

with a complex 'dual variable' ζ. We shall call (*), and its generalizations, the Fourier-Laplace transform of f, restricting the term Fourier transform to the case where the dual variable is real.

Suppose that f is a bounded (measurable) function. If f has compact support, then (*) is analytic on $\mathbf{C}$; if $\operatorname{supp} f \subset [0, \infty)$ then (*) is analytic on $\{\operatorname{Im} \zeta < 0\}$. In either case, one can thus bring in complex variable techniques. This approach can be extended to tempered distributions, multiplied by a suitable exponential if necessary. It is particularly useful in relation to linear partial differential equations with constant coefficients.

10.1 Analytic functions of several complex variables

Let X be an open set in $\mathbf{C}^n$, where $n > 1$, and let $f \in C^1(X)$. Then

$$df = \sum_{j=1}^{n} \frac{\partial f}{\partial z_j}\, dz_j + \sum_{j=1}^{n} \frac{\partial f}{\partial \bar{z}_j}\, d\bar{z}_j,$$

where

$$z = (z_1, \ldots, z_n) = (x_1 + iy_1, \ldots, x_n + iy_n),$$
$$\bar{z} = (\bar{z}_1, \ldots, \bar{z}_n) = (x_1 - iy_1, \ldots, x_n - iy_n),$$

and

$$\frac{\partial}{\partial z_j} = \frac{1}{2}\left(\frac{\partial}{\partial x_j} - i\frac{\partial}{\partial y_j}\right), \quad \frac{\partial}{\partial \bar{z}_j} = \frac{1}{2}\left(\frac{\partial}{\partial x_j} + i\frac{\partial}{\partial y_j}\right), \quad j = 1, \ldots, n.$$

The function f is said to be analytic on X if it satisfies the Cauchy-Riemann equations

$$\frac{\partial f}{\partial \bar{z}_j} = 0, \quad j = 1, \ldots, n. \tag{10.1.1}$$

Let $w \in \mathbf{C}^n$, and let $r = (r_1, \ldots, r_n)$ be an n-tuplet of positive real numbers. The set

$$D(w;r) = \{z\colon |z_j - w_j| < r_j, \quad j = 1, \ldots, r\} \tag{10.1.2}$$

is called a polydisc. Obviously,

$$D(w;r) = D_1 \times \ldots \times D_n, \quad D_j = \{z_j \in \mathbf{C}\colon |z_j - w_j| < r_j\}.$$

Take $w \in X$, and choose r so that the closure of $D(w;r)$ is contained in X. As the functions $z_k \to f(z)$ are analytic functions of a single complex variable when all z_j other than z_k are kept fixed, repeated application of Cauchy's formula yields

$$f(z) = \left(\frac{1}{2\pi \mathrm{i}}\right)^n \int_{\partial D_1} \cdots \int_{\partial D_n} \frac{f(\zeta_1, \ldots, \zeta_n)}{(\zeta_1 - z_1) \ldots (\zeta_n - z_n)} \, \mathrm{d}\zeta_1 \ldots \mathrm{d}\zeta_n \tag{10.1.3}$$

for $z \in D(w;r)$. This shows that $f \in C^\infty(D(w;r))$, since differentiation under the integral sign is legitimate; it also shows that the derivatives $\partial^\alpha f$ satisfy the Cauchy-Riemann equations (10.1.1), and so are analytic in $D(w;r)$. Since X can be covered by such polydiscs, one concludes that, if f is analytic on X, then $f \in C^\infty(X)$, and that its derivatives of all orders are also analytic on X.

As in the case of a single complex variable, the identity (10.1.3) also serves to establish the existence of a power series expansion of f in some neighbourhood of every point $w \in X$. The series on the right hand side of

$$\frac{1}{(\zeta_1 - z_1) \ldots (\zeta_n - z_n)} = \sum_{\alpha \geqslant 0} \frac{(z - w)^\alpha}{(\zeta_1 - w_1) \ldots (\zeta_n - w_n)(\zeta - w)^\alpha}$$

converges uniformly and absolutely on any compact subset of $D(w;r)$. One can therefore substitute it in (10.1.3), and integrate term-by-term. Again, (10.1.3) gives, for every multi-index α,

$$\partial^\alpha f(w) = \left(\frac{1}{2\pi \mathrm{i}}\right)^n \alpha! \int_{\partial D_1} \cdots \int_{\partial D_n} \frac{f(\zeta)}{(\zeta_1 - w_1) \ldots (\zeta_n - w_n)(\zeta - w)^\alpha} \times \mathrm{d}\zeta_1 \ldots \mathrm{d}\zeta_n.$$

Hence

$$f(z) = \sum_{\alpha \geqslant 0} \frac{(z - w)^\alpha}{\alpha!} \partial^\alpha f(w), \tag{10.1.4}$$

with uniform and absolute convergence on compact subsets of $D(w;r)$.

Essentially the only consequence of this basic property of analytic functions which we shall need is the uniqueness of analytic continuation:

Theorem 10.1.1. Let $X \subset \mathbf{C}^n$ be a connected open set. If f is analytic on X, and there is a point $w \in X$ such that $\partial^\alpha f(w) = 0$ for all $\alpha \geqslant 0$, then $f = 0$ on X.

Proof. Let

$$Y = \{z \in X: \partial^\alpha f(z) = 0 \quad \text{for all } \alpha \geqslant 0\}.$$

This set is closed in X, as it is the intersection of a family of closed sets. On the other hand, it follows from (10.1.4) that every point of Y has a polydisc-shaped neighbourhood in Y: hence Y is open in X. By hypothesis, X is connected; so either $Y = X$, or Y is empty. But Y contains the point w, so is not empty; hence $Y = X$, and we are done.

10.2 The Paley–Wiener–Schwartz theorem

If $u \in \mathcal{E}'(\mathbf{R}^n)$, then its Fourier transform is the function

$$\xi \mapsto \langle u(x), \exp(-ix \cdot \xi)\rangle \quad \text{(Theorem 8.4.1).}$$

It is clear that one can replace the real variable ξ by a complex variable, so one can make the following definition:

Definition 10.2.1. If $u \in \mathcal{E}'(\mathbf{R}^n)$, then the function

$$\hat{u}(\zeta) = \langle u(x), e^{-ix \cdot \zeta}\rangle, \quad \zeta \in \mathbf{C}^n, \tag{10.2.1}$$

is called the Fourier–Laplace transform of u.

Note that, if $u \in C_c^\infty(\mathbf{R}^n) = \mathcal{E}'(\mathbf{R}^n) \cap C^\infty(\mathbf{R}^n)$, then (10.2.1) becomes

$$\hat{u}(\zeta) = \int e^{-ix \cdot \zeta} u(x)\, dx, \quad \zeta \in \mathbf{C}^n. \tag{10.2.2}$$

It is clear that, in either case, $\hat{u}(\zeta)$ reduces to the Fourier transform when ζ is real. There are some immediate consequences of the definition:

Theorem 10.2.1. (i) The Fourier–Laplace transform of a member of $\mathcal{E}'(\mathbf{R}^n)$ is an analytic function on $\mathbf{C}^n$.

(ii) If $u \in \mathcal{E}'(\mathbf{R}^n)$ and $\operatorname{supp} u \subset \{|x| \leqslant a\}$, where a is a positive real number, then there are constants $C, N \geqslant 0$ such that

$$|\hat{u}(\zeta)| \leqslant C(1 + |\zeta|)^N e^{a|\operatorname{Im} \zeta|}, \quad \zeta \in \mathbf{C}^n, \tag{10.2.3}$$

where

$$|\zeta| = (|\zeta_1|^2 + \ldots + |\zeta_n|^2)^{1/2}.$$

(iii) If $u \in C_c^\infty(\mathbf{R}^n)$ and $\operatorname{supp} u \subset \{|x| \leqslant a\}$, then there are constants $C_m \geqslant 0$, $m = 0, 1, \ldots$ such that

$$|\hat{u}(\zeta)| \leqslant C_m(1+|\zeta|)^{-m}\, e^{a|\operatorname{Im}\zeta|},\ \zeta \in \mathbf{C}^n, \quad m=0,1,\ldots \tag{10.2.4}$$

Proof. (i) One has $\hat{u} \in C^\infty(\mathbf{C}^n)$, by Corollary 4.1.2, and the $\partial\hat{u}/\partial\zeta_j$, $\partial\hat{u}/\partial\bar{\zeta}_j$, $j=1,\ldots,n$ can be computed by differentiation inside the duality bracket. As $\zeta \mapsto \exp(-ix\cdot\zeta)$ is analytic on $\mathbf{C}^n$, this implies that u satisfies the Cauchy-Riemann equations (10.1.1), and so is an analytic function on $\mathbf{C}^n$.

(ii) Let $\psi \in C^\infty(\mathbf{R})$ be such that $\psi(t)=1$ if $t \geqslant -\frac{1}{2}$ and $\psi(t)=0$ if $t \leqslant -1$, and put

$$\phi_\zeta(x) = \psi(|\zeta|(a-|x|)), \quad x \in \mathbf{R}^n.$$

Then $\phi_\zeta \in C_c^\infty(\mathbf{R}^n)$, and

$$\phi_\zeta = 0 \quad \text{if} \quad |x| \geqslant a+|\zeta|^{-1}, \qquad \phi_\zeta = 1 \quad \text{if} \quad |x| \leqslant a+\tfrac{1}{2}|\zeta|^{-1}.$$

Hence

$$\hat{u}(\zeta) = \langle u(x), \phi_\zeta(x)\, e^{-ix\cdot\zeta}\rangle.$$

This is bounded for $|\zeta| \leqslant 1$, and for $|\zeta| \geqslant 1$ one has $\operatorname{supp}\phi_\zeta \subset \{|x| \leqslant a+1\}$, so that there is a semi-norm estimate for u which gives

$$|\hat{u}(\zeta)| \leqslant C' \sum_{|\alpha| \leqslant N} \sup |D_x^\alpha(\phi_\zeta(x)\, e^{-ix\cdot\zeta})|$$

for some $C', N \geqslant 0$. This implies the estimate (10.2.3), as one can expand each derivative by Leibniz's theorem, and observe that there are constants C_β such that

$$|D_x^\beta \phi_\zeta(x)| \leqslant C_\beta |\zeta|^{|\beta|}, \quad |\zeta| \geqslant 1$$

while

$$|D_x^\gamma\, e^{-ix\cdot\zeta}| = |\zeta^\gamma|\, e^{x\cdot\operatorname{Im}\zeta} \leqslant |\zeta|^{|\gamma|} \exp(|\operatorname{Im}\zeta|(a+|\zeta|^{-1}))$$

for $x \in \operatorname{supp}\phi_\zeta$.

(iii) If $u \in C_c^\infty(\mathbf{R}^n)$, then it follows by repeated partial integration from (10.2.2) that

$$\zeta^\alpha \hat{u}(\zeta) = (-1)^{|\alpha|} \int e^{-ix\cdot\zeta} D^\alpha u(x)\, dx, \quad \zeta \in \mathbf{C}^n$$

for all multi-indices $\alpha \geqslant 0$. So one has, if $\operatorname{supp} u \subset \{|x| \leqslant a\}$,

$$|\zeta^\alpha \hat{u}(\zeta)| \leqslant V_a \sup |D^\alpha u| \sup\{e^{x\cdot\operatorname{Im}\zeta} : x \in \operatorname{supp} u\}$$

where V_a is the measure of the ball $|x| \leqslant a$. Hence

$$|\zeta^\alpha||\hat{u}(\zeta)| \leqslant V_a\, e^{a|\operatorname{Im}\zeta|} \sup |D^\alpha u|, \quad \zeta \in \mathbf{C}^n \tag{10.2.5}$$

for all $\alpha \geqslant 0$, and these inequalities clearly imply (10.2.4).

Conversely, the Fourier-Laplace transforms of distributions with compact support, and of test functions, can be characterized by their asymptotic behaviour as $|\zeta| \to \infty$. This is the content of the Paley-Wiener-Schwartz theorem, which runs as follows:

Theorem 10.2.2. Let a be a positive real number. A function $U(\zeta)$ which is analytic on $\mathbf{C}^n$ is the Fourier-Laplace transform of a distribution $u \in \mathscr{D}'(\mathbf{R}^n)$ supported in $|x| \leqslant a$ if and only if there is an estimate

$$|U(\zeta)| \leqslant C(1+|\zeta|)^N \, \mathrm{e}^{a|\mathrm{Im}\,\zeta|}, \quad \zeta \in \mathbf{C}^n \tag{10.2.6}$$

for some constants $C, N \geqslant 0$.

Furthermore, U is the Fourier-Laplace transform of a function $u \in C_c^\infty(\mathbf{R}^n)$ with $\operatorname{supp} u \subset \{|x| \leqslant a\}$ if and only if there are constants C_m, $m = 0, 1, \ldots$, such that

$$|U(\zeta)| \leqslant C_m(1+|\zeta|)^{-m} \, \mathrm{e}^{a|\mathrm{Im}\,\zeta|}, \quad \zeta \in \mathbf{C}^n \tag{10.2.7}$$

for $m = 0, 1, \ldots$.

Proof. The necessity of the conditions follows from Theorem 10.2.1, so we have only to prove sufficiency. We begin with the C_c^∞ case.

For $\zeta = \xi \in \mathbf{R}^n$, the inequality (10.2.7), with $m = n+1$, shows that $U(\xi) \in L_1(\mathbf{R}^n)$. So one obtains a continuous function u by setting

$$u(x) = (2\pi)^{-n} \int \mathrm{e}^{\mathrm{i}x\cdot\xi} U(\xi)\,\mathrm{d}\xi, \quad x \in \mathbf{R}^n. \tag{10.2.8}$$

If $\alpha > 0$ is a multi-index, then $\xi^\alpha U(\xi) \in L_1(\mathbf{R}^n)$, by (10.2.7) with $m = |\alpha| + n + 1$. Hence one can differentiate under the integral sign in (10.2.8), and conclude that $u \in C^\infty(\mathbf{R}^n)$.

We next show that $\operatorname{supp} u \subset \{|x| \leqslant a\}$. Take $m = n+1$ in (10.2.7). As U is analytic on $\mathbf{C}^n$, one can apply Cauchy's theorem to each variable $\zeta_1, \ldots, \zeta_n$ in turn to shift the integration into the complex domain, and thus replace (10.2.8) by

$$u(x) = (2\pi)^{-n} \int_{\mathrm{Im}\,\zeta = \eta} \mathrm{e}^{\mathrm{i}x\cdot\zeta} U(\zeta)\,\mathrm{d}\zeta, \quad x \in \mathbf{R}^n,$$

where η may be any point in $\mathbf{R}^n$. (This can of course be done at one stroke by using the exactness of the n-form $\mathrm{e}^{-\mathrm{i}x\cdot\zeta} U(\zeta)\,\mathrm{d}\zeta$.) So (10.2.7), with $m = n+1$, gives the estimate

$$|u(x)| \leqslant (2\pi)^{-n} C_{n+1} \, \mathrm{e}^{a|\eta| - x\cdot\eta} \int (1+|\xi|)^{-n-1} \,\mathrm{d}\xi = C' \, \mathrm{e}^{a|\eta| - x\cdot\eta}.$$

Put $\eta = tx/|x|$, where $t > 0$; then this becomes

$$|u(x)| \leqslant C' \, \mathrm{e}^{(a-|x|)t}.$$

If $|x| > a$, one can let $t \to \infty$ to obtain $u(x) = 0$, and so one has $\operatorname{supp} u \subset \{|x| \leqslant a\}$.

We thus have $u \in C_c^\infty(\mathbf{R}^n)$. By the Fourier inversion theorem on $\mathscr{S}(\mathbf{R}^n)$ and (10.2.8), it follows that $\hat{u}(\xi) = U(\xi)$ for $\xi \in \mathbf{R}^n$. Furthermore, by Theorem

10.2.1, $\hat{u}(\xi)$ extends to an analytic function on $\mathbf{C}^n$. So one has $u(\zeta)=U(\zeta)$ for all $\zeta\in\mathbf{C}^n$ by the uniqueness of analytic continuation (Theorem 10.1.1), and we are done.

Turning to the distribution case, where U is analytic on $\mathbf{C}^n$ and satisfies (10.2.6), we observe that this estimate with $\zeta=\xi\in\mathbf{R}^n$ shows that $U(\xi)$, as a distribution, is in $\mathscr{S}'(\mathbf{R}^n)$. So $U(\xi)$ is the Fourier transform of some $u\in\mathscr{S}'(\mathbf{R}^n)$. We now regularize this u by setting $u_\varepsilon=\psi_\varepsilon * u$, where $\varepsilon>0$ and

$$\psi\in C_c^\infty(\mathbf{R}^n),\quad \operatorname{supp}\psi\subset\{|x|\leqslant 1\},\quad \int\psi\,dx=1,\quad \psi_\varepsilon(x)=\varepsilon^{-n}\psi(x/\varepsilon).$$

Then $\hat{u}_\varepsilon(\xi)=U(\xi)\hat{\psi}(\varepsilon\xi)$ by Theorem 8.4.2. By Theorem 10.2.1, $\hat{u}_\varepsilon(\xi)$ thus extends to an analytic function on $\mathbf{C}^n$ such that, for $m=0,1,\ldots$

$$|\hat{u}_\epsilon(\zeta)|\leqslant CC_m(1+|\zeta|)^N(1+\epsilon|\zeta|)^{-m}\,e^{(a+\epsilon)|\operatorname{Im}\zeta|},\quad \zeta\in\mathbf{C}^n,$$

in view of (10.2.4) and (10.2.7). For fixed $\epsilon>0$, one can replace m by $m+N$ here, and conclude that $\hat{u}_\varepsilon$ satisfies estimates of the type (10.2.7), with a replaced by $a+\varepsilon$. Hence, by what has been proved already, u_ε is in C_c^∞ and

$$\operatorname{supp}u_\varepsilon\subset\{|x|\leqslant a+\varepsilon\}. \tag{10.2.9}$$

As $u_\varepsilon\to u$ when $\varepsilon\to 0$, it follows from (10.2.9) that $\operatorname{supp}u\subset\{|x|\leqslant a\}$. Thus also $u\in\mathscr{E}'(\mathbf{R}^n)$, hence $\hat{u}(\xi)$ extends to an analytic function $\hat{u}(\zeta)$ on $\mathbf{C}^n$ by Theorem 10.2.1, and as $\hat{u}(\xi)=U(\xi)$, the uniqueness of analytic continuation implies that $\hat{u}(\zeta)=U(\zeta)$ on $\mathbf{C}^n$. This completes the proof of the theorem.

The estimate (10.2.3), which has played a crucial role in this section, can be sharpened. Let $u\in\mathscr{E}'(\mathbf{R}^n)$, and let ϵ be a positive real number. By regularizing the characteristic function of the 2ϵ-neighbourhood of the support of u, one obtains a function $\psi_\epsilon(x)\in C_c^\infty(\mathbf{R}^n)$ such that $0\leqslant\psi_\epsilon\leqslant 1$, with $\psi_\epsilon(x)=1$ when $d(x,\operatorname{supp}u)\leqslant\epsilon$ and $\psi_\epsilon(x)=0$ when $d(x,\operatorname{supp}u)\geqslant 3\epsilon$, also, for any multi-index α,

$$|D^\alpha\psi_\epsilon|\leqslant C_\alpha\epsilon^{-|\alpha|},$$

where C_α is independent of ϵ. Taking $\phi_\zeta(x)=\psi_\epsilon(x)$ with $\epsilon=1/|\zeta|$ in part (ii) of the proof of Theorem 10.2.1, one finds

$$|\hat{u}(\zeta)|\leqslant C(1+|\zeta|)^N\,e^{p(\operatorname{Im}\zeta)}$$

instead of (10.2.3); here

$$p(\eta)=\sup\{x\cdot\eta : x\in\operatorname{supp}u\},\quad \eta\in\mathbf{R}^n.$$

This is called the support function of the set supp u, and is obviously determined by the convex hull of this set. Likewise, Theorem 10.2.2 can be sharpened so as to yield a relation between the asymptotic behaviour of $\log|U(\zeta)|$ as $|\mathrm{Im}\,\zeta|\to\infty$, and the convex hull of the support of the distribution whose Fourier-Laplace transform is U. (See J. L. Lions, Supports dans la transformation de Laplace, *J. Anal. Math.* 2 (1952-53) 369-80, and L. Gårding, Support functions of plurisubharmonic functions, *J. Math. Mech.* (*Indiana J.*) 17 (1967) 225-40.)

10.3 An application to evolution operators

Distributions with compact support are not the only ones for which a Fourier-Laplace transform can be defined. For instance, suppose that f is a (measurable) function on $\mathbf{R}^n$, and that there are real numbers T and c such that $\operatorname{supp} f\subset\{x_n\geqslant T\}$, and $f\exp(cx_n)\in L_1(\mathbf{R}^n)$. Let $\zeta=(\xi',\zeta_n)$, where $\xi'\in\mathbf{R}^{n-1}$ and $\zeta_n\in\mathbf{C}$. Then it is not difficult to show that

$$\hat{f}(\zeta)=\int e^{-ix\cdot\zeta}f(x)\,dx=\int e^{\eta_n x_n}\,e^{-ix\cdot\xi}f(x)\,dx$$

converges if $\mathrm{Im}\,\zeta_n<c$, and that $\zeta_n\mapsto\hat{f}(\zeta)$ is then analytic. One can generalize this, by considering the subspace of $\mathscr{D}'(\mathbf{R}^n)$ which consists of distributions u supported in some half space $\{x_n\geqslant T\}$, and such that $u\exp(cx_n)\in\mathscr{S}'(\mathbf{R}^n)$ for some $c\in\mathbf{R}$. The Fourier-Laplace transform of u can be defined by

$$\hat{u}(\zeta)=\mathscr{F}(u(x)\,e^{\eta_n x_n})(\xi), \tag{10.3.1}$$

where $\zeta=(\xi',\zeta_n)$ with $\xi'\in\mathbf{R}^{n-1}$ and $\zeta_n=\xi_n+i\eta_n$, and $\eta_n<c$. We shall not develop this theory here, but consider, instead, an application of the underlying principle.

Throughout this subsection, we write $x\in\mathbf{R}^n$ and $\xi\in\mathbf{R}^n$ as (x',x_n) and (ξ',ξ_n), respectively, and set

$$H_n=\{x\in\mathbf{R}^n\colon x_n\geqslant 0\}.$$

Definition 10.3.1. A linear differential operator with constant coefficients, defined on $\mathbf{R}^n$, is called an evolution operator with respect to H_n if it has a fundamental solution whose support is contained in H_n.

To see what the definition implies, take an operator $P(D)$ which is of evolution type, and let E be a fundamental solution of P supported in H_n. If $f\in\mathscr{E}'(\mathbf{R}^n)$ and $u=E*f$, then $Pu=f$ and

$$\operatorname{supp} u\subset\operatorname{supp} f+H_n.$$

Hence

$$Pu = f, \quad u = 0 \quad \text{if} \quad x_n < \inf\{y_n : y \in \operatorname{supp} f\}.$$

If one thinks of x' as a point of 'space', and of x_n as 'time', then this becomes a causality property of P. A fundamental solution of P supported in H_n can thus be called a causal fundamental solution. But it must be borne in mind that the property of the support of u which has just been derived does not necessarily determine u uniquely, if $n > 1$. This will be discussed below.

In the next theorem, a class of evolution operators

$$P(D) = \sum a_\alpha D^\alpha$$

is characterized by an algebraic condition on the polynomial

$$P(\zeta) = \sum a_\alpha \zeta^\alpha$$

which is just the symbol of P.

Theorem 10.3.1. Let $P(\zeta)$, where $\zeta \in \mathbf{C}^n$, be a polynomial which has the following properties:

(i) there is a real number c such that $P(\xi', \zeta_n) \neq 0$ if $\xi' \in \mathbf{R}^{n-1}$ and $\operatorname{Im} \zeta_n < c$;

(ii) the coefficient of the highest power of ζ_n in P is constant.

Then $P(D)$ is an evolution operator with respect to H_n.

Proof. By (ii),

$$P(\xi', \zeta_n) = A(\zeta_n^k + \zeta_n^{k-1} P_1(\xi') + \ldots + P_k(\xi')),$$

where $A \neq 0$ is a complex number, the P_j are polynomials, and the integer k does not exceed the order of P. We may assume that $k \geq 1$, as the hypothesis (ii) renders the case $k = 0$ trivial. So

$$P(\xi', \zeta_n) = A(\zeta_n - \lambda_1) \ldots (\zeta_n - \lambda_k),$$

where the λ_j are functions of ξ'. By hypothesis (i), one has $\operatorname{Im} \lambda_j \geq c$ for $\xi' \in \mathbf{R}^{n-1}$ and $j = 1, \ldots, k$. Since

$$|\zeta_n - \lambda_j| \geq |\operatorname{Im} \lambda_j - \operatorname{Im} \zeta_n|$$

it follows that

$$|P(\xi', \zeta_n)| \geq |A| (C - \operatorname{Im} \zeta_n)^k > 0 \quad \text{if} \quad \operatorname{Im} \zeta_n < c. \tag{10.3.2}$$

So $\xi \mapsto 1/P(\xi', \xi_n + i\eta_n)$ is in $\mathscr{S}'(\mathbf{R}^n)$ for fixed $\eta_n < c$.

We can therefore define a distribution E by setting

$$E = e^{-\eta_n x_n} \mathscr{F}^{-1}(1/P(\xi', \xi_n + i\eta_n)) \tag{10.3.3}$$

where $\eta_n < c$. (Note that $E \in \mathscr{D}'(\mathbf{R}^n)$, but that it is not necessarily tempered. The apparent dependence on η_n will be dealt with presently.) Let $\psi \in C_c^\infty(\mathbf{R}^n)$. It follows from (10.3.3) that

$$\langle E\,\mathrm{e}^{\eta_n x_n}, \check{\psi}\rangle = (2\pi)^{-n}\int \frac{\hat{\psi}(\xi)}{P(\xi', \xi_n + \mathrm{i}\eta_n)}\,\mathrm{d}\xi.$$

We can now set $\psi = \phi\exp(\eta_n x_n)$, so that $\phi \in C_c^\infty(\mathbf{R}^n)$,

$$\langle E\,\mathrm{e}^{\eta_n x_n}, \check{\psi}\rangle = \langle E, \phi\rangle,$$

and

$$\hat{\psi}(\xi) = \int \mathrm{e}^{\eta_n x_n - \mathrm{i}x\cdot\xi}\phi(x)\,\mathrm{d}x = \hat{\phi}(\xi', \xi_n + \mathrm{i}\eta_n) \tag{10.3.4}$$

where $\hat{\phi}$ is the Fourier-Laplace transform of ϕ. It is clear that one can take ϕ to be any member of $C_c^\infty(\mathbf{R}^n)$; so

$$\langle E, \check{\phi}\rangle = (2\pi)^{-n}\int \frac{\hat{\phi}(\xi', \xi_n + \mathrm{i}\eta_n)}{P(\xi', \xi_n + \mathrm{i}\eta_n)}\,\mathrm{d}\xi,\ \phi \in C_c^\infty(\mathbf{R}^n) \tag{10.3.5}$$

provided that $\eta_n < c$. By Theorem 10.2.1, $\hat{\phi}$ is analytic and satisfies the estimates (10.2.4), where one can take $m = n+1$, for example. It is then a simple exercise to deduce from Cauchy's theorem that the second member of (10.3.5) is independent of η_n, as long as $\eta_n < c$.

The distribution which we have constructed is a fundamental solution of P. Indeed, if $\phi \in C_c^\infty(\mathbf{R}^n)$ then

$$\langle PE, \check{\phi}\rangle = \langle E, P(-D)\check{\phi}\rangle = \langle E, (P\phi)\check{}\,\rangle,$$

and as $(P\phi)\hat{} = P(\zeta)\hat{\phi}(\zeta)$, it follows from (10.3.5) that

$$\langle PE, \check{\phi}\rangle = (2\pi)^{-n}\int \hat{\phi}(\xi', \xi_n + \mathrm{i}\eta_n)\,\mathrm{d}\xi.$$

It again follows from Theorem 10.2.1 and Cauchy's theorem that this can be replaced by

$$\langle PE, \check{\phi}\rangle = (2\pi)^{-n}\int \hat{\phi}(\xi)\,\mathrm{d}\xi = \phi(0) = \check{\phi}(0),$$

where we have finally used Fourier's theorem for $\mathscr{S}(\mathbf{R}^n)$, and so $PE = \delta$.

Finally, we show that $\operatorname{supp} E \subset H_n$. Suppose that ϕ in (10.3.5) is supported in $x_n > 0$, and take $\eta_n < \min(0, c)$. Then it follows from (10.3.4), and the identities derived from it by repeated partial integration, that

$$|\hat{\phi}(\xi', \xi_n + \mathrm{i}\eta_n)| \leqslant C(1 + |\xi'| + |\xi_n + \mathrm{i}\eta_n|)^{-n-1}$$

for some constant C. We also have the estimate (10.3.2). Hence (10.3.5) gives

$$|\langle E, \check{\phi}\rangle| \leqslant \frac{C}{(2\pi)^n |A|(c-\eta_n)^k}\int \frac{\mathrm{d}\xi}{(1 + |\xi'| + |\xi_n + \mathrm{i}\eta_n|)^{n+1}}$$

One can now let $\eta_n \to -\infty$, and conclude that $\langle E, \phi\rangle = 0$ if ϕ is supported in

$\{x_n > 0\}$. Since $\check{\phi}(x) = \phi(-x)$, it follows easily that $E = 0$ on $\{x_n < 0\}$, and so the proof is complete.

We shall now consider two examples. It will be assumed that $n > 1$, as both conditions of the theorem are fulfilled trivially when $n = 1$. The first example is

$$P(\xi) = \mathrm{i}\xi_n + \omega|\xi'|^2, \tag{10.3.6}$$

where ω is a complex number. Then (ii) holds and $P(\xi', \zeta_n)$ has the unique zero $\zeta_n = \mathrm{i}\omega|\xi'|^2$. So (i) holds if and only if $\mathrm{Re}\,\omega \geqslant 0$. This includes the heat operator ($\omega = 1$) and the Schrödinger operator ($\omega = \mathrm{i}$). Using Exercise 8.9, it is not difficult to compute the fundamental solution (10.3.5) in this case. For the heat operator, it is just (5.4.12), in a different notation. For complex ω, with $\mathrm{Re}\,\omega \geqslant 0$, see Exercise 10.2.

This fundamental solution is not uniquely determined by the property that it vanishes for $x_n < 0$. To see this, consider [cf. 3, pp. 121-2]

$$u(x) = \int_{\mathrm{i}\tau-\infty}^{\mathrm{i}\tau+\infty} \mathrm{e}^{-\psi(x,s)}\,\mathrm{d}s \tag{10.3.7}$$

where $\tau > 0$, and

$$\psi = \mathrm{i}x_n s + \mathrm{i}x_n(s/\mathrm{i}\omega)^{1/2} + (s/\mathrm{i})^\rho.$$

Here, ρ is a real number, $\frac{1}{2} < \rho < 1$, and

$$(s/\mathrm{i})^\rho = |s|^\rho\,\mathrm{e}^{\mathrm{i}\rho(\theta-\frac{1}{2}\pi)} \quad \text{where} \quad s = |s|\,\mathrm{e}^{\mathrm{i}\theta}, 0 < \theta < \pi;$$

$(s/\mathrm{i}\omega)^{1/2}$ has a fixed determination in the upper half plane. As the term $(s/\mathrm{i})^\rho$ dominates ψ for large $|s|$, it is a straightforward exercise to prove that u is independent of the choice of $\tau > 0$, and that

$$u \in C^\infty(\mathbf{R}^n), Pu = 0, \operatorname{supp} u \subset H_n \tag{10.3.8}$$

Note that this implies that, for any multi-index α, one has $D^\alpha u \to 0$ as $x_n \to 0+$.

Furthermore, one has $u \not\equiv 0$ in every neighbourhood of the origin, so that the causal fundamental solution E fails to be unique in any half space $\{x: x_n < T, T > 0\}$. For if $x_1 = 0$, then (10.3.7) gives

$$u(0, x_n)\,\mathrm{e}^{-\tau x_n} = \int_{-\infty}^{\infty} \exp\left(-\mathrm{i}\sigma x_n - \left(\frac{\sigma + \mathrm{i}\tau}{\mathrm{i}}\right)^\rho\right)\mathrm{d}\sigma.$$

By Parseval's identity, and since $u = 0$ for $x_n < 0$ by (10.3.8), one thus has

$$\int_0^\infty |u(0, x_n)|^2\,\mathrm{e}^{-2\tau x_n}\,\mathrm{d}x_n = (2\pi)\int_{-\infty}^{\infty} \exp\left(-2\,\mathrm{Re}\left(\frac{\sigma + \mathrm{i}\tau}{\mathrm{i}}\right)^\rho\right)\mathrm{d}\sigma.$$

This gives $u(0, x_n) \in L_2(0, \infty)$; also, since

$$0 < \mathrm{Re}\left(\frac{\sigma + \mathrm{i}\tau}{\mathrm{i}}\right)^\rho \leqslant |(\sigma + \mathrm{i}\tau)^\rho| \leqslant (|\sigma| + \tau)^\rho \leqslant |\sigma|^\rho + \tau^\rho$$

(because $0<\rho<1$), one has

$$\int_0^\infty |u(0,x_n)|^2\, e^{-2\tau x_n}\, dx \geqslant 2\pi\, e^{-2\tau^\rho} \int_{-\infty}^{\infty} e^{-2|\sigma|^\rho}\, d\sigma.$$

But if $u \equiv 0$ on a neighbourhood of the origin, then $u(0,x_n)=0$ on some interval $(0,\epsilon)$, and the inequality would imply that

$$e^{2\epsilon\tau}\, e^{-2\tau^\rho} \leqslant M < \infty, \quad \tau>0,$$

which is absurd.

The second example is the wave operator with lower order terms,

$$P(D)=\sum_{j=1}^{n-1} D_j^2 - D_n^2 + \sum_{j=1}^{n} \omega_j D_j + \omega_0, \tag{10.3.9}$$

where $\omega_0, \omega_1, \ldots, \omega_n$ are complex numbers. We can write its symbol as

$$P(\xi)=|\xi'|^2-\xi_n^2+\omega'\cdot\xi'+\omega_n\xi_n+\omega_0.$$

Condition (ii) of Theorem 10.3.1 is satisfied. The zeros of $\zeta_n \mapsto P(\xi',\zeta_n)$ are

$$\zeta_n=\tfrac{1}{2}\omega_n \pm (|\xi'|^2+\omega'\cdot\xi'+\omega_0^2+\tfrac{1}{4}\omega_n^2)^{1/2}.$$

When $\xi' \in \mathbf{R}^{n-1}$ and $|\xi'|$ is large, this gives

$$\zeta_n=\pm|\xi'|+O(1).$$

Hence $|\mathrm{Im}\,\zeta_n|$ is bounded on $\{\zeta_n\colon P(\xi',\zeta_n)=0\}$, so the condition (ii) of the theorem is also fulfilled. Hence P is an evolution operator with respect to H_n. It is obviously also an evolution operator with respect to $\{x_n \leqslant 0\}$.

In fact, one can get more information from (10.3.5). The principal symbol of P,

$$\sigma_P(\xi)=|\xi'|^2-\xi_n^2,$$

is invariant under the Lorentz group. It is not difficult to deduce from this property of σ_P that, if $\eta \in \mathbf{R}^n$ and

$$\eta_n > |\eta'|, \tag{10.3.10}$$

then there is a $t_0=t_0(\eta)$ such that $P(\xi+it\eta)\neq 0$ for $t<t_0$ and all $\xi\in\mathbf{R}^n$. One can then use Cauchy's theorem to show that (10.3.5) can be replaced by

$$\langle E, \check{\phi}\rangle=(2\pi)^{-n}\int \frac{\hat{\phi}(\xi+it\eta)}{P(\xi+it\eta)}\, d\xi, \quad \phi\in C_c^\infty(\mathbf{R}^n),\ t<t_0.$$

Arguing as in the last part of the proof of Theorem 10.3.1, one concludes that $E=0$ if $x\cdot\eta<0$, provided that η satisfies (10.3.10). As the intersection of the complements of these half spaces is the forward null cone, it follows that

$$\operatorname{supp} E \subset \{x\colon x_n \geqslant |x'|\}. \tag{10.3.11}$$

So the proof of Theorem 7.3.2 applies in this case also. In particular, E is the unique causal fundamental solution of the generalized wave operator (10.3.9).

The difference between these two examples comes from the fact that the hyperplanes x_n = constant are characteristics of the operator with the symbol (10.3.6), but not of the operator (10.3.9). (A characteristic of a differential operator $P(x, D)$ is a hypersurface Σ for which there is a defining function S such that $\sigma_P(x, S'(x)) = 0$ on Σ.) If $P(D)$ is a constant coefficient operator on $\mathbf{R}^n$ for which $\sigma_P(0, 1) \neq 0$, then the hyperplanes x_n = constant are not characteristic, and condition (ii) of Theorem 10.3.1 is obviously satisfied. Analysis of condition (i) then shows that, if it is satisfied, the zeros of $\zeta_n \mapsto \sigma_P(\xi', \zeta_n)$ are real when $\xi' \in \mathbf{R}^{n-1}$. A differential operator with this property is usually called hyperbolic. Hyperbolicity is necessary, but (in general) not sufficient, for the validity of (i); hyperbolic operators satisfying condition (i) are called hyperbolic in the sense of Gårding. Their theory generalizes that of the wave operator to differential operators of higher order; see [3, Ch. 5].

Remark. Condition (ii) in Theorem 10.3.1 can be dropped. In fact, it follows from condition (i) of the theorem, and Lemma 2.1 in the Appendix of [3], that there are a $C > 0$ and a rational number d such that

$$|P(\xi', \zeta_n)| \geqslant C(1 + |\xi'| + |\zeta_n|)^d$$

if $\xi' \in \mathbf{R}^{n-1}$ and $\operatorname{Im} \zeta_n \leqslant c - 1$. It is evident that only minor changes in the proof are required when (10.3.2) is replaced by this estimate.

10.4 The Malgrange-Ehrenpreis theorem

If $P(D)$ is a linear differential operator with constant coefficients, defined on $\mathbf{R}^n$, and E is a fundamental solution of P, then the Fourier transform shows formally that $P(\xi)\hat{E} = 1$. So one can find a (tempered) fundamental solution of P by solving a division problem in $\mathscr{S}'(\mathbf{R}^n)$. This can be done, but is beyond the scope of this book, except when $n = 1$, when a tempered fundamental solution can in fact quite easily be constructed without using Fourier transforms. However, the method by which the causal fundamental solution of Theorem 10.3.1 was derived can be modified so as to yield a proof of the existence of a fundamental solution of any differential operator with constant coefficients.

Theorem 10.4.1 (Malgrange, Ehrenpreis). A linear differential operator with constant coefficients has a fundamental solution.

We shall deduce this from the following lemma:

Lemma 10.4.1. Let

$$P(D) = cD_n^m + P_1(D')D_n^{m-1} + \ldots + P_m(D') \tag{10.4.1}$$

be a linear differential operator with constant coefficients, defined on $\mathbf{R}^n$, with

$c \neq 0$ and $D' = (D_1, \ldots, D_{n-1})$. Then there is an $E \in \mathscr{D}'(\mathbf{R}^n)$ such that $PE = \delta$, that is to say, P has a fundamental solution.

Proof. The symbol of P is

$$P(\xi) = c\xi_n^m + P_1(\xi')\xi_n^{m-1} + \ldots + P_m(\xi') \tag{10.4.2}$$

where the P_j are polynomials. For a fixed point $\bar{\xi}' \in \mathbf{R}^{n-1}$ the polynomial $\zeta_n \to P(\bar{\xi}', \zeta_n)$, where $\zeta_n \in \mathbf{C}$, has m (not necessarily distinct) zeros $\lambda_1, \ldots, \lambda_m$. One can therefore find a real number $\tau(\bar{\xi}')$ such that

$$|\tau(\bar{\xi}')| \leqslant m+1, \; |\tau(\bar{\xi}') - \operatorname{Im} \lambda_j| > 1, \quad j = 1, \ldots, m. \tag{10.4.3}$$

It follows from Rouché's theorem that $\bar{\xi}'$ has a neighbourhood $U(\bar{\xi}') \subset \mathbf{R}^{n-1}$ such that (10.4.3) also holds if $\xi' \in U(\bar{\xi}')$. One may obviously take $U(\bar{\xi}')$ to be a cube, for example,

$$U(\bar{\xi}') = \{\xi' : |\xi_j - \bar{\xi}_j| < \delta(\bar{\xi}), \quad j = 1, \ldots, n\}$$

and assume, in addition, that $\delta \leqslant 1$ for all $\bar{\xi} \in \mathbf{R}^{n-1}$. These cubes are an open covering of $\mathbf{R}^{n-1}$. By a routine argument, the Heine-Borel theorem implies that one can extract a countable, locally finite subcover of $\mathbf{R}^{n-1}$ from this, which we can order as $U_j = U(\bar{\xi}'_{(j)}), j = 1, 2, \ldots$. Next, one can define open sets $\Omega_1, \Omega_2, \ldots$ by

$$\Omega_1 = U_1, \Omega_2 = U_2 \setminus \bar{\Omega}_1, \ldots, \Omega_k = U_k \setminus \sum_{j=1}^{k-1} \bar{\Omega}_j, \ldots.$$

The Ω_j are mutually disjoint, the union of their closures is $\mathbf{R}^{n-1}$, and any compact set meets only a finite number of them. Finally, we set, for $j = 1, 2, \ldots$,

$$\Gamma_j = \{\zeta \in \mathbf{C}^n : \zeta' = \xi' \in \Omega_j, \zeta_n = \xi_n + i\tau(\bar{\xi}'_{(j)}), \xi_n \in \mathbf{R}\}. \tag{10.4.4}$$

By construction, (10.4.3) holds in each U_j. Since

$$|P(\zeta)| = |c| \, |\zeta - \lambda_1| \ldots |\zeta - \lambda_m| \geqslant |c| \, |\operatorname{Im} \zeta - \operatorname{Im} \lambda_1| \ldots |\operatorname{Im} \zeta - \operatorname{Im} \lambda_m|$$

it therefore is clear from (10.4.4) that

$$|P(\zeta)| \geqslant |c|, \zeta \in \Gamma_j, \quad j = 1, 2, \ldots \tag{10.4.5}$$

Let us now define a linear form E on $C_c^\infty(\mathbf{R}^n)$ by setting

$$\langle E, \check{\phi} \rangle = (2\pi)^{-n} \sum_{j=1}^{\infty} \int_{\Gamma_j} \frac{\hat{\phi}(\zeta)}{P(\zeta)} \, \mathrm{d}\xi, \quad \phi \in C_c^\infty(\mathbf{R}^n) \tag{10.4.6}$$

We can also write this as

$$\langle E, \check{\phi} \rangle = (2\pi)^{-n} \int_{\Gamma} \frac{\hat{\phi}(\zeta)}{P(\zeta)} \, \mathrm{d}\zeta, \quad \phi \in C_c^\infty(\mathbf{R}^n); \tag{10.4.7}$$

the chain Γ over which one integrates is known as 'Hörmander's staircase'. Each integral in (10.4.6) is well defined, because of (10.4.5). The sum converges, because of the properties of the Ω_j, and the estimate

$$|\hat{\phi}(\xi', \xi_n + i\tau)| \leqslant C(1 + |\xi'| + |\xi_n + i\tau|)^{-n-1} e^{a|\tau|} \tag{10.4.8}$$

which follows from (10.2.4); moreover, the constants C and a can be taken to be the same for all ϕ supported in a fixed compact set. It therefore follows from (10.4.6) that $E \in \mathscr{D}'(\mathbf{R}^n)$.

It remains to prove that $PE = \delta$. As in the proof of Theorem 10.3.1, one has $\langle PE, \check{\phi}\rangle = \langle E, (P\phi)\check{}\,\rangle$ for $\phi \in C_c^\infty(\mathbf{R}^n)$. Hence (10.4.7) gives

$$\langle PE, \check{\phi}\rangle = (2\pi)^{-n} \int_\Gamma P(\zeta) \frac{\hat{\phi}(\zeta)}{P(\zeta)} \, d\zeta = (2\pi)^{-n} \int_\Gamma \hat{\phi}(\zeta) \, d\zeta,$$

whence, by (10.4.8) and Cauchy's theorem,

$$\langle PE, \check{\phi}\rangle = (2\pi)^{-n} \int_{\mathbf{R}^n} \hat{\phi}(\xi) \, d\xi = \phi(0) = \check{\phi}(0), \qquad \phi \in C_c^\infty(\mathbf{R}^n).$$

This proves the lemma, as one can replace ϕ by $\check{\phi}$.

Proof of Theorem 10.4.1. Let $P(D)$ be a differential operator of order m, defined on $\mathbf{R}^n$, and let $\sigma_P(\xi)$ be its principal symbol. Recall that, if $u \in \mathscr{D}'(\mathbf{R}^n)$, and A is a real, constant, nonsingular $n \times n$ matrix, then the pullback of u by the map $x \mapsto Ax$ is

$$\langle A^*u, \phi\rangle = |\det A|^{-1} \langle u, (A^{-1})^*\phi\rangle, \qquad \phi \in C_c^\infty(\mathbf{R}^n).$$

By Corollary 7.2.1, the derivatives of A^*u can be computed by the chain rule. This gives

$$D(A^*u) = A^*({}^tADu),$$

where $D = (D_1, \ldots, D_n)$ is regarded as a column vector, and ${}^tA$ is the transpose of A. Hence

$$P(D)(A^*u) = A^*(P({}^tAD)u). \tag{10.4.9}$$

The coefficient of ξ_n^m in $P({}^tA\xi)$ is $\sigma_P(\theta)$, where θ is vector whose components are the first row of A. One can therefore clearly choose A such that $\sigma_P(\theta) \neq 0$. Thus the differential operator $P({}^tAD)$ has a fundamental solution $F \in \mathscr{D}'(\mathbf{R}^n)$, by the lemma. It then follows from (10.4.9) that

$$P(D)(A^*F) = A^*\delta = |\det A|^{-1} \delta.$$

Hence $E = |\det A| A^*F$ is a fundamental solution of P, and so we are done.

Note. For a more general construction, and another proof, based on the Hahn-Banach theorem, see [3, Ch. 3], where references to the literature can also be found. The proof given above follows I. Gel'fand and G. Shilov, *Generalized functions*, Vol. 2 (Academic Press, 1968).

Exercises

10.1. Let $\zeta \in \mathbf{C}^n$ and put

$$U(\zeta) = \sum_{k=0}^{\infty} (-1)^k \frac{(\zeta\cdot\zeta)^k t^{2k+1}}{(2k+1)!} = \frac{\sin (\zeta\cdot\zeta)^{1/2} t}{(\zeta\cdot\zeta)^{1/2}},$$

with the obvious convention when $\zeta\cdot\zeta = 0$; here, $t \in \mathbf{R}$ is a parameter, $t \neq 0$. Show that U is the Fourier-Laplace transform of a distribution on $\mathbf{R}^n$ whose support is contained in $\{x \in \mathbf{R}^n\colon |x| \leqslant |t|\}$.

What does this imply in conjunction with Exercise 9.5, when the distributions f and g there have compact support?

10.2. Let

$$P(D) = \mathrm{i}D_n + \omega(D_1^2 + \ldots + D_{n-1}^2)$$

be the differential operator whose symbol is (10.3.6); ω is a complex number, $\operatorname{Re}\omega \geqslant 0$. Show that, if $\operatorname{Re}\omega > 0$, then the fundamental solution (10.3.5) is, for $x_n > 0$,

$$E = \left(\frac{1}{2(\pi\omega x_n)^{1/2}}\right)^{n-1} \exp\left(-\frac{|x'|^2}{4\omega x_n}\right),$$

where $\omega^{1/2}$ is chosen so that $\operatorname{Re}\omega^{1/2} > 0$.

Deduce that, if $\omega = \mathrm{i}$ (so that $-\mathrm{i}P$ is the Schrödinger operator), then

$$E = \left(\frac{1}{2(\pi x_n)^{1/2}}\right)^{n-1} \exp\left(-(n-1)\frac{\pi \mathrm{i}}{4} + \frac{\mathrm{i}|x'|^2}{4x_n}\right) \quad \text{if } x_n > 0.$$

10.3. Prove that the function defined by (10.3.7) has the properties (10.3.8).

10.4. Show that

$$-D_n^2 + (1 + \mathrm{i}D_n)(D_1^2 + \ldots + D_{n-1}^2)$$

is an evolution operator on $\mathbf{R}^n$ with respect to $x_n \geqslant 0$.

10.5. Let

$$P(D) = (D_3^2 - a_1^2 D_1^2 - a_2^2 D_2^2)(D_3^2 - a_2^2 D_1^2 - a_1^2 D_1^2),$$

where a_1 and a_2 are real numbers, neither of which is zero. Show that $P(D) + D_1^k$ is an evolution operator on $\mathbf{R}^3$ with respect to $\{x_3 \geqslant 0\}$ if $k = 0$, 1 or 2, but that condition (i) of Theorem 10.3.1 is not satisfied when $k = 3$.

10.6. Let $\mathscr{S}'^{+}(\mathbf{R}) = \mathscr{S}'(\mathbf{R}) \cap \mathscr{D}'^{+}(\mathbf{R})$ be the space of tempered distributions on $\mathbf{R}$ whose supports are bounded on the left. (For $\mathscr{D}'^{+}(\mathbf{R})$, see section 5.3.) The Fourier-Laplace transform of $u \in \mathscr{S}'^{+}(\mathbf{R})$ can be defined as

$$\hat{u}(\zeta) = \mathscr{F}(u\, \mathrm{e}^{\eta x})(\xi), \quad \zeta = \xi + \mathrm{i}\eta \in \mathbf{C}^-,$$

where $\mathbf{C}^- = \{\zeta \in \mathbf{C}\colon \operatorname{Im}\zeta < 0\}$. Show that $(Du)\hat{}\,(\zeta) = \zeta\hat{u}(\zeta)$, and use this and Theorem 8.3.1 to prove that $\hat{u}$ is analytic on $\mathbf{C}^-$, and satisfies

$$|\hat{u}(\zeta)| \leqslant C(|\operatorname{Im}\zeta|^{-1} + |\operatorname{Im}\zeta|^{-M-1})|\zeta|^N \mathrm{e}^{b \operatorname{Im}\zeta}, \quad \zeta \in \mathbf{C}^-,$$

for some constants $C, M, N \geqslant 0$ and $b \in \mathbf{R}$.

Prove that $\hat{u}(\zeta)$ converges to the Fourier transform of u in $\mathscr{S}'(\mathbf{R})$ when $\operatorname{Im}\zeta \to 0-$.

Show also that

$$\hat{u}(\zeta) = \langle u(x), \psi(x)\, \mathrm{e}^{-\mathrm{i}\zeta x}\rangle, \quad \zeta \in \mathbf{C}^-,$$

where $\psi \in C^\infty(\mathbf{R})$ is a cut-off function whose support is bounded on the left, and which is equal to unity on a neighbourhood of the support of u.

10.7. Show that if u and v are members of $\mathscr{S}'^+(\mathbf{R})$, then so is $u * v$, and that $(u * v)\hat{} = \hat{u}\hat{v}$. What does this imply if $u * v = 0$?

10.8. Let $U(\zeta)$ be a function which is analytic on $\mathbf{C}^-$, and such that

$$|U(\zeta)| \leqslant C(|\mathrm{Im}\,\zeta|^{-1} + |\mathrm{Im}\,\zeta|^{-M-1})|\zeta|^N \mathrm{e}^{b\,\mathrm{Im}\,\zeta}, \quad \zeta \in \mathbf{C}^-,$$

for some constants $C, M, N \geqslant 0$ and $b \in \mathbf{R}$. Show that

$$u = \frac{1}{2\pi}(1 + \mathrm{i}D)^{N+2} \int_{\mathrm{i}\tau - \infty}^{\mathrm{i}\tau + \infty} \frac{U(\zeta)}{(1 + \mathrm{i}\zeta)^{N+2}} \mathrm{e}^{\mathrm{i}\zeta x}\, \mathrm{d}\zeta.$$

where $\tau > 0$, is defined independently of τ, and is a member of $\mathscr{S}'^+(\mathbf{R})$ such that $\operatorname{supp} u \subset [b, \infty)$ and $\hat{u}(\zeta) = U(\zeta)$ on $\mathbf{C}^-$.

11 THE CALCULUS OF WAVEFRONT SETS

In previous chapters, various operations on distributions have been defined and upper bounds proven for the singular supports of the resulting distributions. Certain of these operations could only be defined in very limited cases for example multiplication was only defined when the singular supports of the distributions were disjoint. Whilst it is a theorem of Schwartz that a general operation of multiplication cannot be defined for distributions, we shall see in this chapter that by introducing the notion of the direction of a singularity that multiplication can sometimes be defined for distributions which have non-disjoint singular supports. The crucial new notion required is that of wavefront set — the set of directions of singularities of a distribution. As well as enabling us to define operations such as multiplication, pull-back and restriction for distributions, it will also allow us to compute sharper bounds on the singular supports of push-forwards and convolutions. We finish by applying these techniques to the propagation of singularities for the three dimensional wave equation.

We refer the reader whose appetite has been whetted to Hörmander's comprehensive account of micro-local analysis, [**10**].

11.1 Definitions

The Fourier transform of a compactly supported smooth function is a Schwartz function and that of a compactly supported distribution is an analytic function. We therefore define the directions of singularity of a compactly supported distribution to be the directions in which its Fourier transform is not rapidly decreasing.

We need to make precise the notion of direction. As we want our singularity sets to be closed, we require their complements to be open conic neighbourhoods.

Definition 11.1.1. We shall say $\Gamma \subset \mathbf{R}^n \setminus \{0\}$ is conic if $\xi \in \Gamma$ implies that $\lambda\xi \in \Gamma$ for all $\lambda > 0$. A conic neighbourhood of a point shall be an open conic set containing it.

It will be useful to have a notation for something decaying as $|\xi|$ at infinity without bounds on behaviour at zero, so we let

$$\langle \xi \rangle = (1 + |\xi|^2)^{1/2}. \tag{11.1.1}$$

Definition 11.1.2. Let $u \in \mathcal{E}'(\mathbf{R}^n)$ then the direction ξ_0 is not in $\Sigma(u) \subset \mathbf{R}^n \setminus \{0\}$, the frequency set of u, if for all N, there exist C_N such that

$$|\hat{u}(\xi)| \leq C_N \langle \xi \rangle^{-N},$$

for ξ in a fixed conic neighbourhood of ξ_0.

Lemma 11.1.1. Let $u \in \mathcal{E}'(\mathbf{R}^n)$ then $u \in C_c^\infty(\mathbf{R}^n)$ if and only if $\Sigma(u) = \emptyset$.

Proof. Suppose $\Sigma(u) = \emptyset$ then using the compactness of the unit sphere, we can cover $\mathbf{R}^n \setminus \{0\}$ by a finite number of cones in each of which $\hat{u}$ is rapidly decreasing. So there exists constants such that

$$|\hat{u}(\xi)| \leq C_N \langle \xi \rangle^{-N},$$

for all ξ. Now by the Fourier Inversion Theorem (Lemma 8.4.2)

$$u = \left(\frac{1}{2\pi}\right)^n \int e^{ix\cdot\xi} \hat{u}(\xi) d\xi.$$

So as this integral is uniformly convergent independent of x, we have that u is continuous. Applying a differential operator is equivalent to multiplying the Fourier transform by a polynomial so we have that all the derivatives of u are continuous also.

The frequency set of u expresses the direction of singularities without expressing their locations. To obtain the locations, we localize by taking smooth cut-offs.

Definition 11.1.3. Let $u \in \mathcal{D}'(\mathbf{R}^n)$ then we shall say

$$(x_0, \xi_0) \in \mathbf{R}^n \times (\mathbf{R}^n \setminus \{0\})$$

is not in the wavefront set, $\mathrm{WF}(u)$, of u if and only if there exists $\phi \in C^\infty(\mathbf{R}^n)$ such that $\phi(x_0) \neq 0$ and $\xi_0 \notin \Sigma(\phi u)$. For X an open set in $\mathbf{R}^n$, we make the same definition for $u \in \mathcal{D}'(X)$ with $\phi \in C_c^\infty(X)$ to obtain $\mathrm{WF}(u) \subset U \times \{\mathbf{R}^n \setminus \{0\}\}$.

An important point is that multiplying by a compactly supported smooth function can only decrease the frequency set — this means that it is enough to consider cutting off only by functions which are one near the point in question.

Lemma 11.1.2. Let $u \in \mathcal{E}'(\mathbf{R}^n)$ and $\phi \in C_c^\infty(\mathbf{R}^n)$ then

$$\Sigma(\phi u) \subseteq \Sigma(u).$$

Proof. Suppose $\xi_0 \neq 0$ is not in $\Sigma(u)$ then there is an open cone Γ containing ξ_0 which is disjoint from $\Sigma(u)$, in which

$$|\hat{u}| \leq C_N \langle \xi \rangle^{-N}, \ \forall N.$$

Let Γ' be a smaller open cone containing ξ_0. It is sufficient to show that $\widehat{\phi u}$ is rapidly decreasing in Γ'.

Now, as ϕ, u are compactly supported, we have

$$\widehat{\phi u}(\xi) = \left(\frac{1}{2\pi}\right)^n \int \hat{\phi}(\eta)\hat{u}(\xi - \eta)\mathrm{d}\eta,$$

and $\hat{\phi}$ is Schwartz and $|\hat{u}| \leq C(1 + |\xi|)^M$ where M is the order of u as a distribution, using Theorem 10.2.1.

We divide the convolution into two pieces, according to whether ξ is near or far from $\xi - \eta$. Choosing ϵ sufficiently small, we have for $|\eta| < \epsilon|\xi|$, and $\xi \in \Gamma'$ that $\xi - \eta \in \Gamma$. We have that

$$\left| \int_{|\eta|>\epsilon|\xi|} \hat{\phi}(\eta)\hat{u}(\xi - \eta)\mathrm{d}\eta \right| \leq C_N \sup_{|\eta|\geq\epsilon|\xi|} \left(\langle\eta\rangle^{n+1-N+M}\right) \langle\xi\rangle^M \int_{|\eta|\geq\epsilon|\xi|} \langle\eta\rangle^{-n-1}\mathrm{d}\eta,$$

for any N, which is rapidly decreasing in ξ.

For the rest of the integrand, we have that for $\xi \in \Gamma'$ and for any K, N, it is less than

$$C_{K,N} \int_{|\eta|<\epsilon|\xi|} \langle\eta\rangle^{-N}\langle|\xi - \eta|\rangle^{-K}\mathrm{d}\eta$$

which as ϵ is small is less than

$$C\langle\xi\rangle^{-K}.$$

Combining these estimates the result follows.

The wavefront set of a distribution refines the notion of singular support:

Proposition 11.1.1. Let $u \in \mathcal{D}'(\mathbf{R}^n)$ then

$$\operatorname{singsupp}(u) = \{x : \exists\xi\ (x, \xi) \in \mathrm{WF}(u)\}.$$

Proof. If $x \notin \operatorname{singsupp}(u)$ then picking ϕ smooth and supported close to x, we have that ϕu is smooth and compactly supported. We therefore have that $(x, \xi) \notin \mathrm{WF}(u)$ for all ξ which proves the inclusion in one direction.

Suppose (x, ξ) is not in the wavefront set of u for all $\xi \neq 0$. Using the compactness of the unit ball, we can find $\phi_j(x) \in C_0^\infty(\mathbf{R}^n)$ for $j = 1, \ldots, N$ such that $\phi(x) \neq 0$, $\widehat{\phi_j u}$ is rapidly decreasing in a cone Γ_j and $\cup\Gamma_j = \mathbf{R}^n \setminus \{0\}$.

Applying Lemma 11.1.2, we have that

$$\Sigma(\phi_1\phi_2\ldots\phi_N u) = \emptyset$$

and thus by Lemma 11.1.1 $\phi_1\phi_2\ldots\phi_N u$ is smooth and the result follows.

Example 11.1.1. Let $x = (x', x'')$ be a splitting of the coordinates. Let $f(x'')$ be a smooth function, and let u be defined by

$$\langle u, \phi \rangle = \int \phi(0, x'') f(x'') \mathrm{d}x''.$$

The singular support of u is clearly $x' = 0$. What is the wavefront set? Well if $\psi \in C_0^\infty(\mathbf{R}^n)$ then

$$\langle \psi u, \phi \rangle = \int \phi(0, x'') \psi(0, x'') f(x'') \mathrm{d}x''.$$

We thus have that

$$\widehat{\psi u}(\xi) = C_n \widehat{\psi_0 f}(\xi''),$$

where $\psi_0(x'') = \psi(0, x'')$. Now $\psi_0 f$ is smooth and compactly supported so its Fourier transform is Schwartz as a function of ξ'' but constant as a function of ξ'. We therefore conclude that

$$\mathrm{WF}(u) = \{(0, x'', \xi', 0)\}.$$

An important property of wavefront set is that it is always reduced by the application of differential operators. It is enough to consider differentiation in one variable and multiplication by a smooth function.

Let u be a distribution, f be a smooth function and ϕ a compactly supported smooth function. To show that

$$\mathrm{WF}(fu) \subset \mathrm{WF}(u),$$

it is enough to show that

$$\Sigma(\phi f u) \subset \Sigma(\phi u)$$

for such ϕ. If we let ψ be a compactly supported smooth function which is one on the support of ϕ then

$$(\psi f)\phi = f\phi,$$

and applying Lemma 11.1.2 the result follows.

Now for differentiation, suppose that $(x, \xi) \notin \mathrm{WF}(u)$. Then there exists $\phi \in C_0^\infty(\mathbf{R}^n)$ which is identically one near x such that $\xi \notin \Sigma(\phi u)$. Let $\psi \in C_0^\infty(\mathbf{R}^n)$ be such that $\psi(x) \neq 0$ and is supported in the set where ϕ is identically one. So $\psi\phi = \psi$ and $\xi \notin \Sigma(\psi u)$. Now,

$$\psi D_j u = \psi D_j(\phi u) - \psi(D_j \phi) u.$$

The last term is zero and

$$\Sigma(\psi(D_j(\phi u))) \subset \Sigma(D_j(\phi u)),$$

by Lemma 11.1.2. As

$$\widehat{D_j(\phi u)}(\xi) = \xi_j \widehat{\phi u}(\xi),$$

it follows that

$$\xi \notin \Sigma(\psi D_j u) \subset \Sigma(D_j(\phi u)).$$

So to summarize:

Proposition 11.1.2. Let u be a distribution and $P(x, D)$ a differential operator with smooth coefficients then

$$\mathrm{WF}(Pu) \subset \mathrm{WF}(u).$$

This property of differential operators is sometimes called *micro-locality.*

Example 11.1.2. Let $F(x) = H(x_1)$ that is equal 1 on $x_1 \geq 0$ and 0 otherwise. We have that

$$\frac{\partial F}{\partial x_1} = \delta(x_1). \tag{11.1.2}$$

So we conclude that

$$\{(0, x'', \xi_1, 0)\} \subset \mathrm{WF}(F). \tag{11.1.3}$$

To see this is everything consider $\psi(x_1)\phi(x'')F(x)$ where $x = (x_1, x'')$, with ψ, ϕ smooth functions of compact support. This has Fourier transform,

$$\hat{\phi}(\xi'')\widehat{\phi F}(\xi_1), \tag{11.1.4}$$

which is rapidly decreasing in any cone with $|\xi_1| \geq \epsilon|\xi''|$ as $\hat{\phi}(\xi'')$ is Schwartz.

11.2 Transformations of Wavefront Sets under Elementary Operations

We now wish to study how wavefront sets transform under various operations on distributions. We start with four fundamental operations: exterior product; pull-back by a diffeomorphism; multiplication; and push-forward by a projection. Once these are understood, we will rapidly deduce theorems for push-forwards and pull-backs by general smooth functions and in particular we will be able to understand restrictions. We will also be able to prove a theorem on the transformation of wavefront sets upon application of a general operator in terms of the wavefront set of its Schwartz kernel.

Theorem 11.2.1. If $u \in \mathcal{D}'(\mathbf{R}^n)$ and $v \in \mathcal{D}'(\mathbf{R}^k)$ then $\mathrm{WF}(u \otimes v)$
$\subseteq (\mathrm{WF}(u) \times \mathrm{WF}(v)) \cup (\mathrm{WF}(u) \times (\mathrm{supp}(v) \times \{0\}) \cup ((\mathrm{supp}(u) \times \{0\}) \times \mathrm{WF}(v)).$

Proof. As we are only proving an upper bound, it is sufficient to consider cutting $u \otimes v$ off solely by exterior products of compactly supported smooth functions, that is by functions $\phi(x)\psi(y) \in C_0^\infty(\mathbf{R}^n) \times C_0^\infty(\mathbf{R}^k)$. Now the Fourier transform of $(\phi \otimes \psi)(u \otimes v)$ is just $\widehat{\phi u} \otimes \widehat{\psi v}$. If $\widehat{\phi u}(\xi)$ is rapidly decreasing near direction $\xi_0 \neq 0$ or $\widehat{\psi v}(\eta)$ is rapidly decreasing near $\eta_0 \neq 0$ then clearly $\widehat{\phi u} \otimes \widehat{\psi v}(\xi, \eta)$ is rapidly decreasing near (ξ_0, η_0).

So if $(x_0, y_0, \xi_0, \eta_0) \notin \mathrm{WF}(u) \times \mathrm{WF}(u)$ then it is not in $\mathrm{WF}(u \otimes v)$. This leaves points where either ξ_0 or η_0 is zero. Now $\widehat{\phi u} \otimes \widehat{\psi v}(\xi, \eta)$ will be rapidly decreasing near $(\xi_0, 0)$ if $\widehat{\phi u}(\xi)$ is rapidly decreasing near ξ_0 or ψv is zero. Similarly for $(0, \eta_0)$ which proves the result.

We next look at how wavefront sets transform under push-forwards. Push-forward is the analogue of integrating out variables for a distribution. To motivate the definition, consider $K(x, y)$ an L^1 function of compact support on $\mathbf{R}^n \times \mathbf{R}^k$ then

$$\left\langle \int K(x,y)dy, \phi(x) \right\rangle = \langle K(x,y), \phi(x) \otimes 1 \rangle,$$

with the first pairing on $\mathbf{R}^n$ and the second on $\mathbf{R}^n \times \mathbf{R}^k$.

Definition 11.2.1. Let K be a distribution on $\mathbf{R}^n_x \times \mathbf{R}^k_y$ and suppose the projection, π, from the support of K onto the y coordinate is proper then we define $\pi_*(K)$, the push-forward of K by π, by

$$\langle \pi_*(K), \phi \rangle = \langle K, \phi \otimes \psi \rangle$$

for $\text{supp}(\phi) \subset L$, L compact with $\psi \in C_0^\infty(\mathbf{R}^k)$ identically one on a neighbourhood of

$$\{y : \exists x \in L, (x,y) \in \text{supp}(L)\}.$$

(See Section 5.3 for the definition of proper.) The pairing is independent of the choice of ψ because the difference of any two choices will be zero in a neighbourhood of $\text{supp}(K)$. It is easily checked that the resultant operation will be a distribution and that the operation is continuous in the distribution topology. It is obviously linear.

Proposition 11.2.1. Let K be as in Definition 11.2.1 then

$$\text{WF}(\pi_*(K)) \subset \{(x,\xi) : \exists y, (x,y,\xi,0) \in \text{WF}(K)\}.$$

Proof. As usual we show that if $(x, y, \xi, 0)$ is not in $\text{WF}(K)$ for any y then $(x, \xi) \notin \text{WF}(\pi_*(K))$. Fix such a point (x, ξ) then as the fibre above x will be compact, we can choose a finite number of bump functions ϕ_j such that for each y in the fibre above x, some ϕ_j is identically one near (x, y) and the Fourier transform of $\phi_j K$ is rapidly decreasing near $(\xi, 0)$. This means that we can take a partition of unity and reduce to the case where K is of compact support and has Fourier transform rapidly decreasing near $(\xi, 0)$. Observe that the Fourier transform of $\pi_*(K)$ is equal to

$$\langle \pi_* K, e^{-ix\cdot\xi} \rangle = \langle K, e^{-ix\cdot\xi} \otimes \psi \rangle = \langle K, e^{-ix\cdot\xi} \cdot e^{-iy\cdot 0} \rangle$$

with ψ identically one near the y projection of $\text{supp}(K)$. The result is now clear.

The next operation we consider is pull-back by a diffeomorphism. Recall Theorem 7.1.1, that if X, Y are open subsets in $\mathbf{R}^n$ and f is a diffeomorphism then we can define

$$\langle f^*u, \phi\rangle = \langle u(y), g^*\phi(y)|\det g'(y)|\rangle,\ \ \phi \in C_c^\infty(X),$$

for $u \in \mathcal{D}'(X)$ where $g = f^{-1}$. This extends pull-back to distributions and is sequentially continuous in the distribution topology.

We wish to relate the wavefront set of f^*u to that of u. We fix a point x_0 where we compute the wavefront set of f^*u. Since a pull-back by a translation will only multiply Fourier transforms by a factor of modulus one, we can assume that X, Y are neighbourhoods of zero, $x_0 = 0$, and that $f(0) = 0$. Having done this, we can now write

$$f = h \circ A,$$

where A is an invertible linear map and h has derivative equal to the identity at 0. We study each of these cases separately.

Now if $\phi \in C_c^\infty(X)$ then we have,

$$\widehat{A^*(\phi u)}(\xi) = \langle A^*(\phi u), \mathrm{e}^{-\mathrm{i}x\cdot\xi}\rangle \tag{11.2.1}$$

$$= \langle \phi u, \mathrm{e}^{-\mathrm{i}A^{-1}x\cdot\xi}|\det A^{-1}|\rangle \tag{11.2.2}$$

$$= \langle \phi u, \mathrm{e}^{-\mathrm{i}x\cdot(A^{-1})^t\xi}|\det A^{-1}|\rangle. \tag{11.2.3}$$

Of course we really want to understand $\widehat{\psi A^*(u)}(\xi)$ instead, but pull-back will give a bijection on the space of compactly supported smooth functions so this is not important. So the wavefront set of A^*u above 0 is

$$\{(0, (A^{-1})^t\xi) : (0, \xi) \in \mathrm{WF}(u)\}.$$

To finish we show that pull-back by h does not affect the wavefront set at the origin. Let l be equal to h^{-1}. We have that $l(x) = x + O(x^2)$. We want to show that if ψu has rapidly decreasing Fourier transform in direction θ_0 then so does $\tilde{\psi}h^*u$ for some $\tilde{\psi}$ (by symmetry this is sufficient.) We actually examine $h^*(\psi_1\psi_2 u)$ where ψ_j are to be chosen — we can put $\tilde{\psi} = h^*(\psi_1\psi_2)$.

We have

$$\langle h^*(\psi_1\psi_2 u), \mathrm{e}^{-\mathrm{i}x\cdot\xi}\rangle = \langle \psi_1\psi_2 u, \mathrm{e}^{-\mathrm{i}l(x)\cdot\xi}|\det l'|\rangle \tag{11.2.4}$$

$$= \langle \psi_1 u, \psi_2\mathrm{e}^{-\mathrm{i}l(x)\cdot\xi}|\det l'|\rangle \tag{11.2.5}$$

$$= \langle \widehat{\psi_1 u}(\theta), \int \mathrm{e}^{\mathrm{i}x\cdot\theta}\psi_2(x)\mathrm{e}^{-\mathrm{i}l(x)\cdot\xi}|\det l'|\mathrm{d}x\rangle. \tag{11.2.6}$$

So we must consider the integral,

$$\int\int \widehat{\psi_1 u}(\theta)\mathrm{e}^{\mathrm{i}(x\cdot\theta - x\cdot\xi - r(x)\cdot\xi)}\psi_2(x)|\det l'|\mathrm{d}\theta\mathrm{d}x,$$

where $l(x) = x + r(x)$. We divide this into two pieces according to whether θ is in the subset of the set, Γ, where $\widehat{\psi_1 u}(\theta)$ is rapidly decreasing — we assume that ψ_1 has been chosen to ensure that $\theta_0 \in \Gamma$. Let Γ_1 a conic open subset of Γ containing θ_0 and which has closure contained in Γ. We then have

$$|\widehat{\psi_1 u}(\theta)| \leq C_N \langle\theta\rangle^{-N},$$

for $\theta \in \Gamma_1$.

First consider the integral over $\theta \in \Gamma_1$. Putting $\phi(x,\xi) = x \cdot \xi + r(x) \cdot \xi$. We have that

$$d_x\phi = \xi + r_1(x) \cdot \xi,$$

with r_1 vanishing at $x = 0$. So $d_x\phi$ is non-zero for $|\xi| > 1$, for x small. Putting

$$L = \mathrm{i}|d_x\phi|^{-2} \sum_{j=1}^{n} \frac{\partial \phi}{\partial x_j} \frac{\partial}{\partial x_j},$$

we have

$$L\mathrm{e}^{-\mathrm{i}\phi(x,\xi)} = \mathrm{e}^{-\mathrm{i}\phi(x,\xi)},$$

and L has coefficients which are homogeneous of degree -1 in ξ. So integrating by parts in x, we have

$$\int_{\theta\in\Gamma_1} \int (L^t)^k (\mathrm{e}^{\mathrm{i}x\cdot\theta} \widehat{\psi_1 u}(\theta) \psi_2(x) |\det l'|) \mathrm{e}^{-\mathrm{i}(x\cdot\xi + r(x)\cdot\xi)} \mathrm{d}x\mathrm{d}\theta,$$

for any k. As the coefficients of L^t are of order -1 in ξ and $\widehat{\psi_1 u}(\theta)$ is rapidly decaying in θ, this will decay to order $-k$ in ξ for any k. That is the integral is rapidly decreasing in ξ.

Now for the integral in the complement of Γ_1. Let Γ_2 be an open conic subset of Γ_1 such $\theta_0 \in \Gamma_1$ and $\bar{\Gamma}_2 \subset \Gamma_1$. We have that if x is sufficiently small then $d_x\phi = \xi + r_1(x) \cdot \xi \in \Gamma_2$ for ξ in a small conic neighbourhood of θ_0. So provided we take ψ_2 to have sufficiently small support and ξ in this small neighbourhood, we have that

$$d_x(x \cdot \theta - \phi(x,\xi)) \neq 0.$$

So using homogeneity, it follows that

$$|d_x(x \cdot \theta - \phi(x,\xi))| \geq C(|\xi| + |\theta|).$$

Let

$$M = -\mathrm{i}|d_x(x \cdot \theta - \phi(x,\xi))|^{-2} \sum_{j=1}^{n} \frac{\partial}{\partial x_j}(x \cdot \theta - \phi(x,\xi)) \frac{\partial}{\partial x_j}.$$

We then have

$$M\mathrm{e}^{\mathrm{i}(x\cdot\theta - \phi(x,\xi))} = \mathrm{e}^{\mathrm{i}(x\cdot\theta - \phi(x,\xi))}.$$

So the integral equals

$$\int_{\theta\notin\Gamma_1} \mathrm{e}^{\mathrm{i}x\cdot\theta - \mathrm{i}(x\cdot\xi - r(x)\cdot\xi)} \widehat{\psi_1 u}(\theta) (M^t)^k (\psi_2(x)|\det l'|) \mathrm{d}x\mathrm{d}\theta, \tag{11.2.7}$$

for any k. As the coefficients of M^t are of order -1 in $|\xi| + |\theta|$, we deduce that the integral is rapidly decaying in ξ and our result follows.

So to summarize

Proposition 11.2.2. Let X, Y be open subsets of $\mathbf{R}^n$ and let $f : X \to Y$ be a diffeomorphism. Then if $u \in \mathcal{D}'(Y)$, the wavefront set of f^*u is equal to

$$\{(x, df_x^t\eta) : (f(x), \eta) \in \mathrm{WF}(u)\}.$$

Note that this really says that wavefront set transforms under coordinate transformations as a subset of the cotangent bundle and thus that wavefront set can be defined for a distribution on a manifold as a conic subset of the cotangent bundle.

We next consider multiplication. This is not in general defined. So we work with distributions u_1, u_2 such that

$$\mathrm{WF}(u_1) + \mathrm{WF}(u_2) = \{(x, \xi_1 + \xi_2) : (x, \xi_j) \in \mathrm{WF}(u_j)\}$$

does not contain any points $(x, 0)$. Taking a partition of unity, this means we can restrict to the case of multiplying compactly supported distributions, v_1, v_2, such that

$$0 \notin \Sigma(v_1) + \Sigma(v_2).$$

In this case, the convolution

$$\int \hat{v}_1(\xi - \eta)\hat{v}_2(\eta)\mathrm{d}\eta,$$

converges absolutely. To see this, take $p_1, p_2 \in C^\infty(\mathbf{R}^n - \{0\})$ which are homogeneous of degree zero such that p_j is identically 1 on $\Sigma(v_j)$, and

$$0 \notin \mathrm{supp}(p_1) + \mathrm{supp}(p_2).$$

We can then write the convolution as

$$\int (p_1\hat{v}_1)(\xi - \eta)(p_2\hat{v}_2)(\eta)\mathrm{d}\eta + \int ((1-p_1)\hat{v}_1)(\xi - \eta)(p_2\hat{v}_2)(\eta)\mathrm{d}\eta +$$
$$\int (p_1\hat{v}_1)(\xi - \eta)((1-p_2)\hat{v}_2)(\eta)\mathrm{d}\eta + \int ((1-p_1)\hat{v}_1)(\xi - \eta)((1-p_2)\hat{v}_2)(\eta)\mathrm{d}\eta. \tag{11.2.8}$$

The first term is an integral over a compact subset — if η_j is in the support of the integrand for all j and $|\eta_j|$ tend to ∞, then we have $\xi/|\eta_j| - \eta_j/|\eta_j| \in \mathrm{supp}(p_1)$ and $\eta_j/|\eta_j| \in \mathrm{supp}(p_2)$. Taking an appropriate subsequence, we deduce that there exists η of modulus one such that $-\eta \in \mathrm{supp}(p_1)$ and $\eta \in \mathrm{supp}(p_2)$, which contradicts our choice of p_1 and p_2. This will remain true for ξ in a compact set. For all the other terms, at least one member is rapidly decreasing so the integral converges absolutely.

So we define v_1v_2 to be the inverse Fourier transform of this convolution. We claim that

$$\Sigma(v_1v_2) \subset (\Sigma(v_1) + \Sigma(v_2)) \cup \Sigma(v_1) \cup \Sigma(v_2). \tag{11.2.9}$$

It is enough to establish this with $\Sigma(v_j)$ replaced by $\mathrm{supp}(p_j)$ as we can always pick $\mathrm{supp}(p_j)$ arbitrarily close to $\Sigma(v_j)$. If ξ is not in $\mathrm{supp}(p_1)+\mathrm{supp}(p_2)$ then the first term of (11.2.8) is zero. The final term is the convolution of two rapidly decreasing functions and is therefore rapidly decreasing. The other two terms can be regarded as Fourier transforms of smooth functions times distributions and so no new characteristic directions will be introduced — thus they are rapidly decreasing outside $\mathrm{supp}(p_1)$ and $\mathrm{supp}(p_2)$.

So we conclude, after making arbitrarily small cut-offs that

Proposition 11.2.3. If $u_1, u_2 \in \mathcal{D}'(\mathbf{R}^n)$ are such that $\mathrm{WF}(u_1) + \mathrm{WF}(u_2)$ contains no zero directions then the product u_1u_2 can be defined and

$$\mathrm{WF}(u_1u_2) \subset \mathrm{WF}(u_1) \cup \mathrm{WF}(u_2) \cup (\mathrm{WF}(u_1) + \mathrm{WF}(u_2)). \tag{11.2.10}$$

We made some choices in this definition; we can show that the product is choice independent by showing that it is continuous in a certain sense or more simply by observing that associativity holds and then using a product of two partitions of unity to see that the choice is independent of which one used.

Observe that the condition for the product to exist is precisely that the upper bound for the wavefront set of the product contains no zero directions. We will see this is typical — an operation can be carried out if the natural bound for the wavefront set after the operation has no zero directions.

Before going on to some harder operations, let's look at some examples. Let G be equal to one inside the unit disc in $\mathbf{R}^2$ and zero outside. What is its wavefront set? The singular support is $|x| = 1$, so the projection of the wavefront set is $|x| = 1$. So suppose we take polar coordinates, (y_1, y_2), near a point on $|x| = 1$, with y_1 the distance from the origin and y_2 the angle. In these coordinates, G becomes the function F from Example 11.1.2. So the wavefront set of G is the pull-back of the wavefront set of F. We can now either compute derivatives and apply Proposition 11.2.2 or we can find a geoemetrically invariant way to characterize the wavefront set here. We do the second and leave the first to the reader as an exercise. We can identify the wavefront set of F as the set of covectors which pair with vectors tangent to $y_1 = 0$ to give zero. So the wavefront set of G is the set of covectors which pair with the tangent vectors to the circle to give zero. That is the vectors normal to the circle which are of course the multiples of the radius. So we conclude that

$$\mathrm{WF}(G) = \{(x, \xi) : |x| = 1,\ \xi = \lambda x, \lambda \neq 0\}. \tag{11.2.11}$$

Now suppose we push-forward from $\mathbf{R}^2$ to $\mathbf{R}$ by

$$\pi(x, y) = x. \tag{11.2.12}$$

The projection of the singular support is the interval $[-1,1]$. What does the wavefront set tell us? We have that $\mathrm{WF}(\pi_*(G))$ is contained in the projection of points in the wavefront set where $\xi_2 = 0$. The only such points are $(-1,0,\xi_1,0)$ and $(1,0,\xi_1,0)$. So we conclude that the wavefront set of the push-forward is contained in, $(-1,\xi_1),(-1,\xi_1')$, and thus that the singular support in contained in $\{-1,1\}$ which is much smaller than the projection of the singular support. As wavefront set is only decreased on multiplication by smooth functions, this remains true of $\mathrm{WF}(\pi_*(fG))$ for any smooth function f.

Now consider the distribution $u(x) = H(x_1)\delta(x'')$ using our result on exterior products, and the support of u we conclude that

$$\mathrm{WF}(u) \subset \{(0,0,\xi_1,\xi'')\} \cup \{(x_1,0,0,\xi''), x_1 \geq 0\}. \tag{11.2.13}$$

In fact, as $\frac{\partial u}{\partial x_1} = \delta(x)$, we have that $\mathrm{WF}(u)$ contains the entire fibre above the origin. Also as $\mathrm{WF}(u)$ equals $\delta(x'')$ in $x_1 > 0$ the bound is sharp there too.

11.3 Push-Forwards and Pull-Backs

We can now prove results about push-forwards and pull-backs for quite general smooth functions by decomposing into the elementary operations.

We next discuss pull-backs by proper smooth functions. These are not always defined but given a relation between the wavefront set and the function, we will see that they exist. We commence with restrictions to coordinate planes which can be viewed as pull-back by the inclusion map. So suppose, we have split the coordinates x on $\mathbf{R}^n$, $x = (x', x'')$ with $x' = (x_1, \dots, x_k)$. If $u \in \mathcal{D}'(\mathbf{R}^n)$, we have

$$u_{|x'=0} = \pi_*(u\delta(x')),$$

where $\pi(x', x'') = x'$, and $\delta(x')$ is just the distribution obtained by integrating over $x' = 0$. We saw above that the wavefront set of $\delta(x')$ is $(0, x'', \xi', 0)$. So the product exists provided

$$\mathrm{WF}(u) \cap \{(0, x'', \xi', 0)\} = \emptyset.$$

So we have, using Proposition 11.2.1, that

Proposition 11.3.1. Let $u \in \mathcal{D}'(\mathbf{R}^n)$ be such that $\mathrm{WF}(u) \cap \{(0, x'', \xi', 0)\}$ is empty then the restriction of u to $x' = 0$ exists and we have the wavefront set relation,

$$\mathrm{WF}(u_{|x'=0}) \subset \{(x'', \xi'') : \exists \xi',\ (0, x'', \xi', \xi'') \in \mathrm{WF}(u)\}.$$

Now consider a general smooth map, F, from $U_1 \subset \mathbf{R}^n$ to $U_2 \subset \mathbf{R}^m$. For convenience we assume that F is proper — pull-back by F will then map

$C_c^\infty(U_2)$ to $C_c^\infty(U_1)$ We decompose the map into simpler maps in order to apply the results already proven. Let T_F be the map

$$T_F(x, y) = (x, y + F(x))$$

then

$$F(x) = \pi_y(T_F(x, 0)).$$

So we can regard pull-back as the composition of π_y^*, pull-back by the diffeomorphism T_F and restriction to $y = 0$.

Now pull-back by π_y is just the same as taking the exterior product with 1_x so we define,

$$F^*(u) = (T_F^*(1 \otimes u))_{|y=0},$$

when the restriction exists. Let's compute what this means in terms of wavefront sets and what the wavefront set of F^*u is. The wavefront set of $1 \otimes u$ is just

$$(x, 0, y, \eta), \ (y, \eta) \in \mathrm{WF}(u).$$

Apply Proposition 11.2.2, to deduce that

$$\mathrm{WF}(T_F^*(1 \otimes u)) = T_F^* \, \mathrm{WF}(1 \otimes u). \tag{11.3.1}$$

Now $T_F(x, y) = (x, y + F(x))$, so

$$dT_F = \begin{pmatrix} \mathrm{Id} & 0 \\ dF & \mathrm{Id} \end{pmatrix} \tag{11.3.2}$$

and so

$$dT_F^t = \begin{pmatrix} \mathrm{Id} & dF^t \\ 0 & \mathrm{Id} \end{pmatrix}. \tag{11.3.3}$$

Thus we have,

$$\mathrm{WF}(T_F^*(1 \otimes u)) = \{(x, dF_x^t \eta, y + F(x), \eta) : (y + F(x), \eta) \in \mathrm{WF}(u)\}. \tag{11.3.4}$$

The restriction to $y = 0$ will exist if and only if there are no points of the form $(x, 0, F(x), \eta) \in \mathrm{WF}(T_F^*(1 \otimes u))$. So our pull-back is defined if $(F(x), \eta) \in \mathrm{WF}(u)$ implies that $dF_x^t \eta \neq 0$ and the wavefront set of the pull-back is then contained in

$$\{(x, dF_x^t \eta) : (F(x), \eta) \in \mathrm{WF}(u)\}. \tag{11.3.5}$$

So to summarize:

Proposition 11.3.2. Let U_1, U_2 be open sets in $\mathbf{R}^n$ and let $F : U_1 \to U_2$ be smooth and proper. If $u \in \mathcal{D}'(U_2)$ then we can define F^*u the pull-back of u provided $dF_x^t \eta \neq 0$ for $(F(x), \eta) \in \mathrm{WF}(u)$ and we have

$$\mathrm{WF}(F^*(u)) \subset \{(x, dF_x^t \eta) : (F(x), \eta) \in \mathrm{WF}(u)\}. \tag{11.3.6}$$

The dual operation to pull-back is of course push-forward. This is always defined for distributions of compact support but may not give a map on compactly supported smooth functions. That is the push-forward of a smooth function may not be smooth. We prove:

Proposition 11.3.3. Let U_1, U_2 be open sets in $\mathbf{R}^n$ and let $F : U_1 \to U_2$ be a smooth map. Then we can define

$$\langle F_* u, \phi \rangle = \langle u, F^* \phi \rangle,$$

for $u \in \mathcal{D}'(U_1)$ and $\phi \in C_c^\infty(U_2)$. We then have that $F_* u$ is a distribution of compact support and has wavefront set contained in,

$$\{(y, \eta) : y \in F(\mathrm{supp}(u)), y = F(x), dF_x^t \eta = 0\} \\ \cup \{(y, \eta) : y = F(x), (x, dF_x^t \eta) \in \mathrm{WF}(u)\}. \quad (11.3.7)$$

Proof. That this actually gives us a distribution of compact support follows from the fact that pull-back is a sequentially continuous map on functions of compact support and that the image of a compact set is compact, (see proof of Theorem 7.1.1.)

To get the wavefront set relation, we observe that

$$F_* u = (\pi_x)_* ((u(x) \otimes 1) \delta(y - F(x))),$$

and apply our previous results.

The wavefront set of $\delta(y - F(x))$ is the pull-back of that of $\delta(y)$ under the diffeomorphism,

$$(x, y) \mapsto (x, y - F(x)),$$

and so equals

$$\{(x, -dF_x^t \eta, F(x), \eta)\}. \quad (11.3.8)$$

The wavefront set of $(u(x) \otimes 1)\delta(y - F(x))$ is then

$$(x, \xi - dF_x^t \eta, F(x), \eta),$$

and so when we push-forward in x we obtain $\{(F(x), \eta)\}$ such that

$$(x, dF_x^t \eta) \in \mathrm{WF}(u)$$

or $dF_x^t \eta = 0$, which proves the theorem. Note that if u is zero near $F^{-1}(x)$, that $F^{-1}(x) \cap \mathrm{supp}(u)$ is empty then $x \notin \mathrm{supp}(F_* u)$ so there can be no wavefront set at x.

11.4 Wavefront Sets and Schwartz Kernels

We have proved theorems about how wavefront set tranforms under various operations. Since any operation has a Schwartz kernel, we now look at how the wavefront set of the Schwartz kernel affects the wavefront set of the distribution under the associated operation.

Applying an operator, P, with Schwartz kernel $K \in \mathcal{D}'(\mathbf{R}^{m+n})$ to a function ϕ gives the distribution,

$$\langle P\phi, \psi \rangle = \langle K, \psi \otimes \phi \rangle. \tag{11.4.1}$$

So the operation of applying $K(x, y)$ to $u(y) \in \mathcal{E}'(\mathbf{R}^n)$ can be written as

$$Pu = (\pi_x)_*(K.(1 \otimes u)), \tag{11.4.2}$$

where $\pi_x(x, y) = x$. So the main issue is whether the product $K \cdot (1 \otimes u)$ is well-defined. Now

$$\mathrm{WF}(1 \otimes u) = \{(x, y, 0, \eta) : (y, \eta) \in \mathrm{WF}(u)\}. \tag{11.4.3}$$

So the product is well-defined if $(x, y, 0, -\eta) \in \mathrm{WF}(K)$ implies $(y, \eta) \notin \mathrm{WF}(u)$, by Proposition 11.2.3 and has wavefront set $(x, y, \xi, \eta_1 + \eta_2)$, where $(x, y, \xi, \eta_1) \in \mathrm{WF}(K)$ and $(y, \eta_2) \in \mathrm{WF}(u)$, together with $(x, y, \xi, \eta) \in \mathrm{WF}(K), y \in \mathrm{supp}(u)$ and $(x, y, 0, \eta)$ with $(y, \eta) \in \mathrm{WF}(u)$. Applying Proposition 11.2.1, we deduce that

$$\begin{aligned} &\mathrm{WF}((\pi_x)_*(K.(1 \otimes u))) \subset \\ &\quad \{(x, \xi) : \exists (y, \eta), (x, \xi, y, -\eta) \in \mathrm{WF}(K), (y, \eta) \in (\mathrm{WF}(u) \cup (\mathbf{R}^n \times \{0\})\}. \end{aligned} \tag{11.4.4}$$

If we write (rearranging the order of coordinates)

$$\mathrm{WF}'(K) = \{(x, \xi, y, \eta) : (x, \xi, y, -\eta) \in \mathrm{WF}(K)\}, \tag{11.4.5}$$

and let $\circ$ denote composition of relations, then we can write this as

$$\mathrm{WF}(Pu) \subset \mathrm{WF}'(K) \circ (\mathrm{WF}(u) \cup (\mathbf{R}^n \times \{0\}). \tag{11.4.6}$$

So to summarize:

Theorem 11.4.1. Let P be an operator with Schwartz kernel

$$K \in \mathcal{D}'(\mathbf{R}^m \times \mathbf{R}^n),$$

then P can be applied to $u \in \mathcal{D}'(\mathbf{R}^n)$ provided $(y, n) \in \mathrm{WF}(u)$ implies that there are no points $(x, 0, y, -\eta) \in \mathrm{WF}(K)$. We then have

$$\mathrm{WF}(Pu) \subset \mathrm{WF}'(K) \circ (\mathrm{WF}(u) \cup (\mathbf{R}^n \times \{0\}). \tag{11.4.7}$$

Corollary 11.4.1. If K has wavefront set contained in $\{\xi \neq 0, \eta \neq 0\}$ then K defines maps

$$P : C_c^\infty(\mathbf{R}^m) \to C^\infty(\mathbf{R}^n), \tag{11.4.8}$$

$$P : \mathcal{E}'(\mathbf{R}^m) \to \mathcal{D}'(\mathbf{R}^n). \tag{11.4.9}$$

Convolution can be regarded as a special case of this where

$$K(x, y) = v(x - y).$$

The wavefront set of K can be computed from the pull-back theorem as it is the pull-back of v under

$$(x, y) \mapsto x - y. \tag{11.4.10}$$

So

$$\mathrm{WF}(K) \subset \{(x, x - \alpha, \eta, -\eta) : (\alpha, \eta) \in \mathrm{WF}(v)\}. \tag{11.4.11}$$

So convolution with v will induce a map from compactly supported smooth functions to smooth functions and from compactly supported distributions to distributions by Corollary 11.4.1 (which we already knew.) We also have that

$$\mathrm{WF}(u * v) \subset \{(x, \eta) : \exists \alpha\ (\alpha, \eta) \in \mathrm{WF}(u), (x - \alpha, \eta) \in \mathrm{WF}(u)\} \tag{11.4.12}$$

$$= \{(x + y, \xi) : (x, \xi) \in \mathrm{WF}(u),\ (y, \xi) \in \mathrm{WF}(v)\}, \tag{11.4.13}$$

which is considerably better than the vector sum of the singular supports which is the best that can be done without using wavefront sets. Only points in the wavefront set which are in the same direction add.

Recall from Theorem 8.6.1 that if $P(\partial)$ is elliptic then applying $P(\partial)$ does not decrease the singular support. The same is true for wavefront sets. The proof of Theorem 8.6.1 constructed a distribution, E, of compact support such that

$$P(\partial)E = \delta_0 + r, \tag{11.4.14}$$

with $r \in C_c^\infty(\mathbf{R}^n)$. So the singular support of E is $x = 0$ and by micro-locality we have that

$$\mathrm{WF}(E) = \{(0, \xi)\}. \tag{11.4.15}$$

So we have that

$$\mathrm{WF}(E * u) \subset \mathrm{WF}(u) \tag{11.4.16}$$

and thus

$$\mathrm{WF}(v) = \mathrm{WF}(E * P(D)v) \subset \mathrm{WF}(P(D)v) \subset \mathrm{WF}(v) \tag{11.4.17}$$

for any $v \in \mathcal{D}'(\mathbf{R}^n)$.

11.5 Propagation of Singularities

In this section, we apply the theory developed in this chapter to the wave equation in $\mathbf{R} \times \mathbf{R}^3$. Recall from Section 7.3 that

$$P(\partial) = \partial_4^2 - \partial_1^2 - \partial_2^2 - \partial_3^2, \tag{11.5.1}$$

and that it has fundamental solutions, $E^{\pm}$, where

$$\langle E_+, \phi\rangle = \frac{1}{2\pi}\int \phi(x', \pm|x'|)|x'|^{-1}\mathrm{d}x'. \tag{11.5.2}$$

By micro-locality, $\mathrm{WF}(E^{\pm})$ must contain all points over $x = 0$. What else does it contain? Well away from the origin it is an integral over the cone $x_4 = \pm|x'|$, so applying Proposition 11.2.2 and recalling Example 11.1.1, we deduce that the wavefront set of $E^{\pm}$ is the fibre over the origin together with points

$$(x', \pm|x'|, \lambda(x', \mp|x'|))\ , \lambda \in \mathbf{R} - \{0\}.$$

We apply this to study the possible wavefront sets of solutions of the equation $P(\partial)u = 0$, and more generally the structure of $\mathrm{WF}(Pu) \setminus \mathrm{WF}(u)$.

Proposition 11.5.1. If $u \in \mathcal{D}'(\mathbf{R}^n)$ then

$$\mathrm{WF}(u) \subset \mathrm{WF}(Pu) \cup \{P(\xi) = 0\}. \tag{11.5.3}$$

Proof. Suppose $(y, \eta) \in \mathrm{WF}(u) \setminus \mathrm{WF}(Pu)$. Let $\phi(x) \in C^\infty(\mathbf{R})$ be identically one in $x > y_4 - 1$ and zero in $x < y_4 - 2$. We then have that

$$u = \phi(x_4)u + (1 - \phi)(x_4)u. \tag{11.5.4}$$

Now $\phi(x_4)u = K^+ * (P(\partial)\phi(x_4)u)$ as the convolution exists. As $\phi(x_4)$ is identically one near (y, η) we have that $(y, \eta) \in \mathrm{WF}(\phi(x_4)u) \setminus \mathrm{WF}(P(\partial)\phi(x_4)u)$. So when we convolve with K^+ the wavefront at (y, η) must come from the cone $x_4 = |x'|$ and so must have wavefront contained in $\xi_4 = \pm|\xi'|$. That is $P(\eta) = 0$.

The next question we must answer is which subsets of $P(\xi) = 0$ are possible wavefront sets when $P(\partial)u \in C^\infty(\mathbf{R}^4)$. The answer is that singularities will move in straight lines. When $P(\partial)u$ is not smooth these lines will start and end at points in $\mathrm{WF}(Pu)$.

Theorem 11.5.1. Let $u \in \mathcal{D}'(\mathbf{R}^4)$ then $\mathrm{WF}(u) \setminus \mathrm{WF}(Pu) \subset \{P(\xi) = 0\}$ and is invariant under the flow of the vector field

$$\xi_4\frac{\partial}{\partial x_4} - \xi_1\frac{\partial}{\partial x_1} - \xi_2\frac{\partial}{\partial x_2} - \xi_3\frac{\partial}{\partial x_3}, \tag{11.5.5}$$

in the complement of $\mathrm{WF}(Pu)$.

Note that as $|\xi_4| = \pm|\xi'|$ the integral curves of this vector field are just lines:

$$(x' \mp s\xi'/|\xi'|, x_4 + s, \xi). \tag{11.5.6}$$

As wavefront set is always closed, it is enough to prove a local statement of this theorem as the global result then follows from the fact that the wavefront set will be open and closed in the integral curves. We therefore have only to show that $(x, \xi) \in \mathrm{WF}(u) \setminus \mathrm{WF}(Pu)$ implies that $(x' \mp s\xi'/|\xi'|, x_4 + s, \xi) \in \mathrm{WF}(u)$ for s small.

Now if we let ϕ be a bump function supported near (x, t) and identically one nearer (x, t) then

$$\phi u = K^{\pm} * (P\phi u). \tag{11.5.7}$$

As $(x, \xi) \in \mathrm{WF}(u)$, we conclude that it must come from some point in $\mathrm{WF}(P\phi u)$. It cannot come from the point (x, ξ) as $(x, \xi) \notin \mathrm{WF}(Pu)$ and $x \notin \mathrm{supp}(u - \phi u)$. Let us assume that the support of ϕ is such that

$$(x' \mp s\xi'/|\xi'|, x_4 + s, \xi) \notin \mathrm{WF}(Pu) \text{ for } (x' \mp s\xi'/|\xi'|, x_4 + s) \in \mathrm{supp}(\phi). \tag{11.5.8}$$

So the only way the point at (x, ξ) can arise is through the part of $K^{\pm}$ on the cone interacting with a point of the wavefront set of $P(\phi u)$ on the line $(x' \mp s\xi'/|\xi'|, x_4 + s, \xi)$. This can only happen on the support of the derivative of ϕ. So if we take K^{+}, we have that a point on the support of $d\phi$ of the form $(x' \mp s\xi'/|\xi'|, x_4 + s, \xi)$, with $s < 0$ is in the wavefront set of $P(\phi u)$ and therefore in the wavefront set of u. Now since we can make the support of $d\phi$ arbitrarily small about any point we choose in the ray to x, we deduce that

$$(x' \mp s\xi'/|\xi'|, x_4 + s, \xi) \in \mathrm{WF}(u) \tag{11.5.9}$$

for s small and less than or equal to zero. Applying the same argument with K^{-} gives the corresponding result with $s \geq 0$ and thus the main theorem follows.

We remark that we have really proved a very special case of a very general theorem due to Hörmander on the propagation of singularities of solutions of any partial differential equation with real principal symbol which states that the wavefront set is a union of integral curves of the Hamiltonian vector field of the equation's principal symbol.

Exercises

11.1. Show that if f is a smooth function on $\mathbf{R}^n$ which is supported everywhere then

$$\mathrm{WF}(fu) = \mathrm{WF}(u)$$

for any distribution u.

11.2. Show that the push-forward of the forward fundamental solution of the four-dimensional wave operator by projecting on (x_2, x_3, x_4) is the fundamental solution of the three-dimensional wave operator. Use this to prove propagation of singularities for the three-dimensional operator.

11.3. Show that if $u \in \mathcal{E}'(\mathbf{R}^n)$ and $P(\partial)$ has principal symbol, p_m, then

$$\Sigma(u) \subset \Sigma(Pu) \cup p_m^{-1}(0).$$

11.4. Let f be the function on $\mathbf{R}^2$ which is equal to 1 on $x_1, x_2 \geq 0$ and 0 elsewhere. What is the wavefront set of f ?

11.5. Let $a(\xi)$ be a smooth function on $\mathbf{R}^n$ such that

$$|D_\xi^\alpha a(\xi)| \leq C_\alpha \langle \xi \rangle^{m-\rho|\alpha|},$$

for some $\rho > 0$. Show that the Fourier transform of u is singular only at the origin. Use this to construct a distribution which has wavefront set equal to a single ray.

11.6. Let H be the Heaviside function. Let $f(x,y) = H(y)$ and also $g(x,y) = H(x^2 - y)$ on $\mathbf{R}^2$. Compute the wavefront sets of f, g and fg. Compare the wavefront set of fg with the statement of Proposition 11.2.3.

APPENDIX: TOPOLOGICAL VECTOR SPACES

In this book, the theory of distributions has been developed virtually without explicit reference to topological vector spaces. The purpose of the Appendix is to sketch in some of this background. In particular, it aims to explain how the definition of spaces of test functions by means of semi-norm estimates is related to the general notion of a continuous map between topological spaces.

A *topology* on a set x is a family T of subsets of X with the following properties: (i) the empty set and X itself belong to T; (ii) the union of the members of any subfamily of T is a member of T; (iii) the intersection of a finite number of members of T is a member of T. The pair (X, T) is called a *topological space*; the members of T are called its *open sets*. One sometimes omits explicit reference to T, and speaks of the open sets of a topological space X.

If x is an element ('point') of X, then an open set containing x is called an *open neighbourhood* of x. Any set containing x which also contains an open neighbourhood of x is called a *neighbourhood* of x.

Let T_1 and T_2 be topologies on the same set X. If every member of T_1 is also a member of T_2, then one says that T_1 is *smaller* (or, coarser) than T_2, and that T_2 is *larger* (or, finer) than T_1. Now if S is a family of subsets of X, then there are topologies on X in which all members of S are open sets, for example the so-called discrete topology which consists of all subsets of X. An easy argument shows that the intersection of these (as sets of subsets of X) is again a topology, and that it is the smallest topology containing all the members of S. It is called the topology *generated* by S. Obviously, it consists of unions of finite intersections of members of S. If every member of the topology generated by S is a union of members of S, then S is called a *base* of this topology. Different sets S, or different bases, may generate the same topology.

As an example of this procedure, consider the product of two topological spaces X and Y. Products of sets open in X and sets open in Y are evidently the base of a topology on $X \times Y$. This is called the *product topology*.

Let X and Y be topological spaces. A function (map) f: $X \to Y$ is called *continuous at* $x \in X$ if the inverse image of every open neighbourhood of $f(x)$ is an open neighbourhood of x. It is called *continuous on* X if it is continuous at every point of X. It is obvious that this will be the case if and only if the inverse image of every open set of Y is open in X.

In order to relate this to sequential continuity one must first introduce convergence. A sequence $(x_j)_{1 \leq j < \infty}$ of points of a topological space X is said to *converge* to $x \in X$ as $j \to \infty$ if every neighbourhood of x contains all but a finite number of the x_j. It is now an easy exercise to show that if $f: X \to Y$ is continuous, and $\lim_{j \to \infty} x_j = x$, then $\lim_{j \to \infty} f(x_j) = f(x)$.

A topological space X is said to be *separated*, or *Hausdorff*, if it satisfies the Hausdorff separation axiom: for any two distinct points x and y of X, there is a neighbourhood U of x and a neighbourhood V of y such that $U \cap V$ is empty. In a Hausdorff space, the limit of a convergent sequence is unique.

A *metric space* is a set X equipped with a *distance function.* This is a function $d: X \times X \to \mathbf{R}$ such that, for any $x, y \in X$,

$$d(x,y) = d(y,x), \quad d(x,y) \geq 0, \quad d(x,y) = 0 \Leftrightarrow x = y \tag{1}$$

and

$$d(x,y) \leq d(x,z) + d(z,y), \quad x,y,z \in X \tag{2}$$

(This is the *triangle inequality.*) The *open balls* $\{x: d(x,y) < r\}$, where $y \in X$ and $r > 0$, are then the base of a Hausdorff topology on X, and it is customary to equip a metric space with this topology. For a sequence of points $(x_j)_{1 \leq j < \infty}$ one then has $\lim_{j \to \infty} x_j = x$ if and only if $d(x, x_j) \to 0$ as $j \to \infty$. By the triangle inequality, it follows that then $d(x_j, x_k) \to 0$ as $j, k \to \infty$. Any sequence with this property is called a *Cauchy sequence.* A metric space is called *complete* if every Cauchy sequence converges. For example, $\mathbf{R}$ and $\mathbf{C}$ become complete metric spaces if one takes $d(x,y) = |x - y|$, and the metric gives the usual topology in each case. But, with the same distance function, the set of rational numbers is an incomplete metric space.

A topological space is called *metrizable* if its topology can be derived from a distance function, which is of course not determined uniquely by the topology.

We can now consider vector spaces. Let X be a vector space over $\mathbf{C}$. (One could equally well take a vector space over $\mathbf{R}$, or over any topological field.) It becomes a *topological vector space* when it is equipped with a topology which is compatible with its vector structure. This means that the operations of vector addition, and of multiplication by a complex number, are continuous maps $X \times X \to X$ and $\mathbf{C} \times X \to X$, respectively; here $X \times X$ and $\mathbf{C} \times X$ are given their product topologies. To specify such a topology, it is sufficient to give the neighbourhood system of the zero element $0 \in X$, as the neighbourhood system of any other element of X is obtained from this by translation. In fact, it is enough to give a *base of the neighbourhoods of* 0. This is a family B of neighbourhoods of 0 such that every neighbourhood of 0 contains a member of B.

A set $G \subset X$ is then open if and only if, for every $x \in G$, there is a $U \subset B$ such that $x + U \subset G$, where $x + U = \{y \in X: y - x \in U\}$.

A topological vector space is called *locally convex* if there is a base of neighbourhoods of 0 consisting of sets which are convex, balanced, and absorbing. *Convex* means that $tx + (1-t)y \in U$ if $x, y \in U$ and $0 \leqslant t \leqslant 1$; *balanced*, that $cx \in U$ if $x \in U$ and $|c| \leqslant 1$; *absorbing*, that if $x \in X$, then there is a $t > 0$ such that $x \in tU = \{ty: y \in U\}$. All the vector spaces met with in the theory of distributions are locally convex.

There is an important procedure for constructing locally convex vector space topologies. A *semi-norm* on a vector space X is a function $p: X \to \mathbf{R}$ such that

$$p(cx) = |c| p(x), \quad c \in \mathbf{C}, \quad x \in X, \tag{3}$$

and

$$p(x + y) \leqslant p(x) + p(y), \quad x, y \in X. \tag{4}$$

It follows from (3) that $p(0) = 0$, and hence from (3) and (4) that $0 \leqslant p(x) + p(-x) = 2p(x)$; hence

$$p(x) \geqslant 0, \quad x \in X. \tag{5}$$

A family P of semi-norms on X is called *separating* if, for each $x \in X$, there is a $p \in P$ such that $p(x) \neq 0$. Given such a family, let B be the collection of sets of the form

$$\{x: p(x) < \epsilon, \quad p \in P, \quad \epsilon > 0\}, \tag{6}$$

and of the intersections of a finite number of such sets. Then it is not difficult to prove that B is a base of the neighbourhoods of 0 of a locally convex vector space topology on X. This is called the topology *generated by the semi-norms* belonging to P.

Conversely, let X be a locally convex topological vector space. If U is a convex, balanced, and absorbing neighbourhood of 0, then

$$p_U(x) = \inf \{t > 0: x \in tU\} \tag{7}$$

is called the *Minkowski functional* of U. It is an easy exercise to show that this is a semi-norm, and to deduce that the topology of X can be generated by a family of semi-norms.

Let us consider some examples. If P consists of a single element, then the separation axiom requires that (5) be supplemented by

$$p(x) = 0 \quad \text{if and only if } x = 0. \tag{8}$$

One then calls p a *norm*, and (X, p) a *normed space.* Norms are usually written as $\|x\|$, with a subscript when appropriate. A normed space is metrizable. Indeed, if one puts

$$d(x, y) = \|x - y\|, \quad x, y \in X \tag{9}$$

then it follows from (4), (5) and (7) that d satisfies (1) and (2). Moreover, d gives a *translation invariant* metric, that is to say one has $d(x+z, y+z) = d(x, y)$ for all x, y and z in X. It is obvious that the topology of the metric space (X, d) is the same as the topology generated by the norm. A normed space which is complete with respect to the metric (9) is called a *Banach space.*

For example, let $K \subset \mathbf{R}^n$ be a compact set, and N a nonnegative integer. Then

$$C_C^N(K) = \{\phi \in C^N(\mathbf{R}^n): \operatorname{supp} \phi \subset K\} \tag{10}$$

becomes a Banach space when it is equipped with the norm

$$\phi \mapsto \sum_{|\alpha| \leq N} \sup |\partial^\alpha \phi|; \tag{11}$$

the proof is a simple exercise on uniform convergence.

Again, let P be a countable family of semi-norms, say $P = (p_1, p_2, \ldots)$. If this is to be separating, one must have $x = 0$ if $p_k(x) = 0$ for all $k = 1, 2, \ldots$. The topology generated by such a family of semi-norms is metrizable. For instance, one can put

$$d(x, y) = \sum_{k=1}^{\infty} \frac{1}{2^k} \frac{p_k(x-y)}{1 + p_k(x-y)}, \quad x, y \in X. \tag{12}$$

It is not difficult to prove that this is a distance function, satisfying (1) and (2), and that the topology of the metric space (X, d) coincides with the topology generated by P.

Conversely, it can be shown that the topology of a locally convex space which is metrizable, with a translation invariant metric, can be generated by a countable family of semi-norms. In practice, it is usually easier to work with the semi-norms, rather than with the metric.

A *Fréchet space* is a locally convex topological vector space which is metrizable and complete. In terms of a countable base of semi-norms $(p_1, p_2, \ldots)$ generating the topology, this means that, if $p_m(x_j - x_k) \to 0$ as $j, k \to \infty$, for $m = 1, 2, \ldots$, then there is an element x of the space such that $p_m(x - x_j) \to 0$ as $j \to \infty$, for each $m = 1, 2, \ldots$.

Banach spaces and Fréchet spaces are useful because certain important consequences of Baire's theorem hold in them. One of these was used as a (crucial) step in the proof of Theorem 6.1.1.

The space $\mathscr{S}(\mathbf{R}^n)$ of rapidly decreasing test functions is a Fréchet space, with the topology generated by the semi-norms

$$\phi \to \sup |x^\beta \partial^\alpha \phi|, \quad \alpha \geq 0$$

Likewise, the space $C^\infty(X)$, where $X \subset \mathbf{R}^n$ is an open set, is a Fréchet space. Let $K_j, j = 1, 2, \ldots$ be a sequence of compact sets whose union is X, such that $K_j \subset K_{j+1}$ for all j. The semi-norms generating the topology can be taken to be

$$\phi \mapsto \sup\{|\partial^\alpha \phi(x)| : x \in K_j\}, \quad \alpha \geqslant 0, j = 1, 2, \ldots$$

and one gets the same topology for all admissible sequences of compact sets (K_j). Finally, let $K \subset \mathbf{R}^n$ be a compact set, and

$$C_c^\infty(K) = \bigcap_{N=0}^{\infty} C_c^N(K) = \{\phi \in C_c^\infty(\mathbf{R}^n) \colon \operatorname{supp} \phi \subset K\}. \tag{13}$$

This becomes a Fréchet space when it is given the topology generated by the norms (11), with $N = 0, 1, \ldots$. (The members of a family of semi-norms generating a topology can of course be norms; such a family is manifestly a separating one.) In each of these examples, the completeness proof is an easy exercise which is left to the reader.

Let us now return to the general case of a vector space equipped with a topology generated by a family of semi-norms, and obtain a criterion for the continuity of a linear form. We shall deduce this from the following, more general, proposition:

Proposition A.1. Let X be a locally convex topological vector space, and let P be a family of semi-norms which generates the topology of X. Then a semi-norm q on X is continuous if and only if there is a finite set of semi-norms $p_1, \ldots, p_N$, all belonging to P, and a constant C, such that

$$q(x) \leqslant C \max(p_1(x), \ldots, p_N(x)), \quad x \in X. \tag{14}$$

Proof. If q is continuous, then $q^{-1}(-1, 1)$ is an open neighbourhood of 0, and so contains a finite intersection of sets of the form (6). So there are semi-norms $p_1, \ldots, p_N$, all belonging to P, and positive numbers $\epsilon_1, \ldots, \epsilon_N$, such that $q(y) < 1$ if $p_j(y) < \epsilon_j$ for $j = 1, \ldots, N$. Let x be any element of X; setting $y = tx$, where $t > 0$, one sees that

$$tq(x) < 1 \quad \text{if} \quad tp_j(x) < \epsilon_j, \quad j = 1, \ldots, N. \tag{*}$$

If $\max(p_1(x), \ldots, p_N(x)) > 0$, this gives (14) with $C = 2/\min(\epsilon_1, \ldots, \epsilon_N)$ if one takes $t = \frac{1}{2}\min(\epsilon_1, \ldots, \epsilon_N)/\max(p_1(x), \ldots, p_N(x))$. If $p_j(x) = 0$ for all $j = 1, \ldots, N$, then (*) holds for all $t > 0$, whence $q(x) = 0$, and so (14) holds trivially, with the same C.

Conversely, it is obvious that q is continuous at 0 when (14) holds. Now it readily follows from (4), with p replaced by q, that

$$q(x - y) \leqslant |q(x) - q(y)|, \quad x, y \in X.$$

Hence a semi-norm which is continuous at 0 is continuous on X, and the proposition is proved.

It is a corollary of this proposition that the topology of X is also generated by the family of all continuous semi-norms.

Now let u be a linear form on X. Then $x \mapsto |\langle u, x\rangle|$ is a semi-norm, and so Proposition A.1 gives

Proposition A.2. Let X be as in Proposition A.1. Then a linear form u on X is continuous if and only if there is a finite set of semi-norms $p_1, \ldots, p_N$, all belonging to P, and a constant C, such that

$$|\langle u, x\rangle| \leqslant C \max (p_1(x), \ldots, p_N(x)), \quad x \in X. \tag{15}$$

This statement can often be simplified, as follows:

Proposition A.2′. Suppose that the family P of semi-norms generating the topology of X has the following property: if p_1 and p_2 belong to P, then there is a $p \in P$ such that $p_1(x) \leqslant p(x)$ and $p_2(x) \leqslant p(x)$ for all $x \in X$. Then a linear form u on X is continuous if and only if there is a $p \in P$ and a constant C such that

$$|\langle u, x\rangle| \leqslant Cp(x), \quad x \in X. \tag{16}$$

Proof. Sufficiency is obvious. Again, Proposition A.2 implies the inequality (15) when u is continuous. Easy induction shows that there is a $p \in P$ such that $\max (p_1, \ldots, p_N) \leqslant p$, and so (15) implies (16), and we are done.

When the hypothesis of Proposition A.2′ holds, then Proposition A.1 can be simplified in the same manner.

The set of all continuous linear forms on a locally convex topological vector space is a vector space, which is called the *dual* of X and denoted by X'. The functions $u \mapsto |\langle u, x\rangle|$, where $x \in X$ is fixed, are clearly semi-norms on X', and they are a separating family. So they generate a topology on X', which is called the *weak* topology*. This is not metrizable when X is a Fréchet space, for instance, so that we now have an example of a locally convex topological vector space that is not metrizable.

Proposition A.2′ shows that, if $X \subset \mathbf{R}^n$ is an open set, and $C^\infty(X)$ is topologized in the manner described above, then Definition 3.1.1 is equivalent to defining $\mathscr{E}'(X)$ as the dual of $C^\infty(X)$. Similarly, Definition 8.3.1, and the inequalities (8.3.1), imply that the dual of $\mathscr{S}(\mathbf{R}^n)$ consists of the extensions of tempered distributions to linear forms on $\mathscr{S}(\mathbf{R}^n)$. But more work is needed to deal with the definition of distributions by semi-norm estimates in Chapter 1.

Let $X \subset \mathbf{R}^n$ be an open set. For each compact set $K \subset X$, let $C_c^\infty(K)$ be the vector space (13), with the topology generated by the semi-norms (11), where $N = 0, 1, \ldots$. It is not difficult to prove that the balanced convex subsets of $C_c^\infty(X)$ which meet every $C_c^\infty(K)$ in an open set are a base for the neighbourhoods of 0 of a locally convex topology on $C_c^\infty(X)$. We shall equip $C_c^\infty(X)$ with this topology; it is then called the *inductive limit* of the $C_c^\infty(K)$. This topology does not have a countable base of the neighbourhoods of 0. It therefore follows from a theorem in Functional Analysis that it is not metrizable.

It is evident that the inclusions $C_c^\infty(K) \subset C_c^\infty(X)$ are continuous maps, for every compact subset K of X. (In fact, the inductive limit topology is the largest topology on $C_c^\infty(X)$ with this property.) So if u is a continuous linear form on $C_c^\infty(X)$, then its restriction to any $C_c^\infty(K)$ is continuous. By Proposition A.2 there is therefore a $C \geqslant 0$ and an integer $N \geqslant 0$, both depending on K, such that

$$|\langle u, \phi\rangle| \leqslant C \sum_{|\alpha| \leqslant N} \sup |\partial^\alpha \phi|, \quad \phi \in C_c^\infty(K).$$

Conversely, suppose that u is a linear form on $C_c^\infty(X)$ which satisfies such semi-norm estimates for every compact set $K \subset X$. Then it follows from Proposition A.2 that the restriction of u to any $C_c^\infty(K)$ is continuous, and this evidently implies that u is continuous on $C_c^\infty(X)$. We have thus recovered Definition 1.3.1, and can conclude that $\mathscr{D}'(X)$ is the dual of $C_c^\infty(X)$ in which this space is given the inductive limit topology. The convergence of distributions, according to Definition 1.5.1, is now seen to be convergence in the weak * topology of $\mathscr{D}'(X)$.

We can also elucidate the definition of convergent sequences of test functions in section 1.3. Let $(\phi_j)_{1 \leqslant j < \infty}$ be a sequence of functions of class $C_c^\infty(X)$ which converges in the topology of $C_c^\infty(X)$. As one can subtract the limit of the sequence from each ϕ_j, it is no loss of generality to assume that $\phi_j \to 0$ in $C_c^\infty(X)$ as $j \to \infty$. Suppose that the supports of the ϕ_j are not in a fixed compact subset of X. Let K_j, where $j = 1, 2, \ldots$ be a sequence of compact subsets of X, such that $K_j \subset K_{j+1}$ for all j, whose union is X. Going over to a subsequence, if necessary, one can find points $x^{(1)} \in K_1, x^{(j)} \in K_j \setminus K_{j-1}, j = 2, \ldots$, such that $\phi_j(x^{(j)}) \neq 0$. (This sequence of points clearly cannot have an accumulation point in X.) Put (with $K_0 = \emptyset$)

$$p(\phi) = \sum_{j=1}^{\infty} \sup \{|\phi(x)/\phi_j(x^{(j)})| : x \in K_j \setminus K_{j-1}\}, \quad \phi \in C_c^\infty(X).$$

If K is a compact subset of X, and $\operatorname{supp} \phi \subset K$, then $p(\phi)$ can be estimated by a constant multiple of $\sup |\phi|$. Hence p, restricted to $C_c^\infty(K)$, is a continuous semi-norm (Proposition A.1). The definition of the inductive limit topology then implies that p is a continuous semi-norm on $C_c^\infty(X)$. Hence $\phi_j \to 0$ in $C_c^\infty(X)$ implies that $p(\phi_j) \to 0$ as $j \to \infty$; but, by construction, one has $p(\phi_j) \geqslant 1$ for all j. This contradiction shows that, if a sequence (ϕ_j) converges in $C_c^\infty(X)$, then the supports of the ϕ_j are contained in a fixed compact set K. The sequence thus converges in $C_c^\infty(K)$. The converse is obvious, and so we see now that Definition 1.3.2 is equivalent to convergence in the topology of $C_c^\infty(X)$.

Finally, let X and Y be locally convex topological vector spaces, and let $f: X \to Y$ be a linear map (a vector space homomorphism). It is easy to adapt the proofs of Propositions A.1 and A.2 so as to obtain necessary and sufficient conditions for the continuity of f in terms of semi-norm estimates. This is how Definitions 2.8.1 and 6.2.1 are derived.

BIBLIOGRAPHY

[1] Dieudonné, J. *Foundations of modern analysis*, Academic Press, New York and London, 1960. (New and revised edition: 1969.)

[2] Gel'fand, I. and Shilov, G. *Generalized functions*, Vol. 1, Academic Press, New York and London, 1964.

[3] Hörmander, L. *Linear partial differential operators*, Springer, Berlin–Göttingen–Heidelberg, 1963.

[4] Rudin, W. *Functional analysis*, McGraw Hill, New York, 1973. (Also TMH edition, Delhi, 1974.)

[5] Rudin, W. *Real and complex analysis*, 2nd ed. McGraw Hill, New York, 1973. (Also TMH edition, Delhi, 1973.)

[6] Schwartz, L. *Théorie des distributions*, Hermann, Paris, 1966.

[7] Trèves, F. *Topological vector spaces, distributions and kernels*, Academic Press, New York and London, 1967.

[8] Yosida, K. *Functional analysis*, 5th ed. Springer, Berlin–Heidelberg–New York, 1978.

[9] Zemanian, A. *Distribution theory and transform analysis*, McGraw Hill, New York, 1965.

Additional references

[10] L. Hörmander. *The Analysis of Linear Partial Differential Operators*, Vol. 1 to 4, Springer Verlag, Berlin, 1985.

[11] F. Riesz, B. Sz.-Nagy. *Functional Analysis*, Dover, New York, 1990.

NOTATION

General

R: real line; $\mathbf{R}^n$: product of n copies of **R**
C: complex plane; $\mathbf{C}^n$: product of n copies of **C**
$\mathbf{S}^m$: m-dimensional unit sphere
Z: rational integers; $\mathbf{Z}^n$: product of n copies of **Z**
$x = (x_1^2 + \ldots + x_n^2)^{1/2}$: length of vector $x = (x_1, \ldots, x_n) \in \mathbf{R}^n$
$x \cdot y = x_1 y_1 + \ldots x_n y_n$: inner product in $\mathbf{R}^n$
$[a, b]$, (a, b), $[a, b)$: closed, respectively open, respectively half open interval
$\mathbf{R}^+ = (0, \infty)$: open half line
$\bar{A}$: closure of $A \subset \mathbf{R}^n$; ∂A: boundary of $A \subset \mathbf{R}^n$
$d(x, A)$: distance of point $x \in \mathbf{R}^n$ from a set $A \subset \mathbf{R}^n$
$A \setminus B = \{x \in A : x \notin B\}$: difference of two sets
$A \pm B$: vector sum (difference) of two subsets of $\mathbf{R}^n$
$\partial_j = \partial/\partial x_j, j = 1, \ldots, n$
$\partial^\alpha = \partial_1^{\alpha_1} \ldots \partial_n^{\alpha_n}$, α a multi-index
$|\alpha| = \alpha_1 + \ldots + \alpha_n$: length (order) of multi-index α
$\alpha! = \alpha_1! \ldots \alpha_n!$
$x^\alpha = x_1^{\alpha_1} \ldots x_n^{\alpha_n}$, where $x \in \mathbf{R}^n$ and α is a multi-index
$D_j = (-\sqrt{-1})\partial_j, j = 1, \ldots, n$
$D^\alpha = D_1^{\alpha_1} \ldots D_n^{\alpha_n} = (-\sqrt{-1})^{|\alpha|} \partial^\alpha$, α a multi-index
τ_h: translation by a vector $h \in \mathbf{R}^n$
$\check{}$: reflection in the origin
$\otimes$: tensor product
$*$: convolution
supp f: support of a function or distribution f
Δ: the Laplacian $\partial_1^2 + \ldots + \partial_n^2$ in $\mathbf{R}^n$
f^*: pullback map induced by a map f
${}^tP$: adjoint of the differential operator P
$H(x)$: Heaviside function

$x_+ = xH(x)$, where $x \in \mathbf{R}$

δ, δ_y: Dirac distribution

$\hat{f}$, also $\mathscr{F}f$: Fourier transform (of function or distribution)

$\hat{f}(\zeta), \zeta \in \mathbf{C}^n$: Fourier–Laplace transform (of function or distribution)

$\langle \, , \, \rangle$: pairing of a distribution and a test function

$(\, , \,)_H$: inner product in Hilbert space H

$\| \ \|$: (also with appropriate subscript) norm in a Banach or Hilbert space, semi-norm

Spaces of functions and distributions

X usually stands for an open set in $\mathbf{R}^n$

$C^k(X), C^\infty(X)$	p. 4
$C_c^k(X), C_c^\infty(X)$	p. 5
$C_0^0(\mathbf{R}^n), C_0^k(\mathbf{R}^n)$	pp 124, 125
$\mathscr{D}'(X)$	p. 7
$\mathscr{D}'^m(X)$	p. 9
$\mathscr{D}'^+(\mathbf{R}), \mathscr{D}'^-(\mathbf{R})$	p. 57
$\mathscr{E}'(X)$	p. 34
$H_s(\mathbf{R}^n)$	p. 120
$L_1(X), L_1(\mathbf{R}^n)$	p. 114
$L_1^{\mathrm{loc}}(X)$	p. 8
$L_2(X), L_2(\mathbf{R}^n)$	p. 114
$\mathcal{O}_M(\mathbf{R}^n)$	p. 102
$\mathcal{O}_c(\mathbf{R}^n)$	p. 103
$\mathscr{S}(\mathbf{R}^n)$	p. 93
$\mathscr{S}'(\mathbf{R}^n)$	p. 97

INDEX